ROUTLEDGE LIBR
POLITICAL GE

Volume 14

THE GEOGRAPHY OF STATE POLICIES

THE GEOGRAPHY OF STATE POLICIES

J. R. V. PRESCOTT

LONDON AND NEW YORK

First published in 1968

This edition first published in 2015
by Routledge
2 Park Square, Milton Park, Abingdon, Oxon, OX14 4RN

and by Routledge
711 Third Avenue, New York, NY 10017

Routledge is an imprint of the Taylor & Francis Group, an informa business

British Library Cataloguing in Publication Data
A catalogue record for this book is available from the British Library

ISBN: 978-1-138-80830-0 (Set)
eISBN: 978-1-315-74725-5 (Set)
ISBN: 978-1-138-81595-7 (Volume 14)
eISBN: 978-1-315-74647-0 (Volume 14)
Pb ISBN: 978-1-138-81597-1 (Volume 14)

Publisher's Note
The publisher has gone to great lengths to ensure the quality of this reprint but points out that some imperfections in the original copies may be apparent.

Disclaimer
The publisher has made every effort to trace copyright holders and would welcome correspondence from those they have been unable to trace.

Printed and bound by CPI Group (UK) Ltd, Croydon, CR0 4YY

THE GEOGRAPHY OF STATE POLICIES

J. R. V. Prescott

HUTCHINSON UNIVERSITY LIBRARY
LONDON

HUTCHINSON & CO (*Publishers*) LTD
178–202 Great Portland Street, London W1

London Melbourne Sydney
Auckland Bombay Toronto
Johannesburg New York

First published 1968

This book has been set in Times, printed in Great Britain on Smooth Wove paper by Anchor Press, and bound by Wm. Brendon, both of Tiptree, Essex

09 088860 X (cased)
09 088861 8 (paper)

CONTENTS

MAPS

PREFACE

I would like to thank Professor W. G. East for his encouragement in writing this book. I am also grateful to Professor J. Andrews for stimulating discussions on various aspects of national policy; to my wife for preparing the index; to Mr P. Singleton of the Baillieu Library in the University of Melbourne for untiring assistance in the collection of references; and to Mr H. J. Collier for drawing the maps so well.

1968 J. R. V. PRESCOTT

I

AIMS AND CONCEPTS

Hartshorne's definition of political geography as being 'the study of areal differences and similarities in political character as an interrelated part of the total complex of areal differences and similarities' (1954, p. 178),[1] has gained general acceptance. This measure of agreement stems from the recognition that political authority may be a more important influence on the way in which a person lives and works than the rest of the environment in which he lives. Since it is the government of any independent state which wields political authority, and makes the laws which are an important part of the total environment in which the majority of the world's population lives, it follows that the attention of the political geographer should be focussed, at least in part, on the governments of states. This is not new, but too often in studies of political geography authors refer to France or Britain without reference to the particular government in power in the particular country. Yet a British Conservative or Labour Government may have different geographical viewpoints, and may exert quite different influences on the political geography of the British Isles and overseas countries with which Britain has connections, or for which Britain has responsibility.

Although there has been a long awareness of this, it does not seem to have prompted the logical outcome of more attention on individual governments. Van Valkenburg (1939) included a chapter on the nature of government, because of its influence on relations with other states. Sprout (1962, chapter 6) examined the

[1] For a list of references see end of each chapter.

proposition that the form of a government affects its foreign policies and international capabilities, but came to no final conclusion on the best form. Moodie noted that the 'hallmark of the State is its sovereign power, with the corollary of allegiance by its inhabitants. In practice that sovereignity is exercised by a central government' (1947, p. 54). It is the government which carries out state functions and if there is no government there can be no state. It therefore follows that if any area lies outside the authority of the government it is not part of the state. This view may be legally incorrect but it reflects geographical realities. For example, during the period 1960–2 the Congolese Government in Leopoldville had no authority in Katanga, and since 1959 there have been some parts of South Vietnam which have been persistently beyond the control of authorities in Saigon. Clearly if we seek a more functional political geography our attention must be increasingly focussed on governments.

It is appropriate at this point to refer to an editorial by Cohen (1966), which was entitled 'a geography of policy', in which he suggests that a study of public and private policy provides common ground for economic and political geographers. There is no question of Cohen seeking to create a new branch of geography, he is merely recognising the fact that it is through policy decisions and their implementation that governments and private firms influence geography. Once again it is possible to find precedents for these views, although most earlier comments have been general rather than specific. For example, Pounds (1963, p. *v*) indicates that 'the geographical nature, the policy and the power of the state' are the three main themes running through his book. The book follows a systematic treatment and the geographical analysis of policy is the theme least satisfactorily explored. The discussion of national strategy by Jones (1954) is relevant for the political geographer interested in policy. He endorses Hilsman's plea for political scientists to become more policy oriented. For Jones national strategy or policy is the second ray of a country's power fix, the first is resources. There is little point in assessing national power unless the purpose of which the power is to be deployed is also known. This point has been stressed by Sprout:

> . . . elaborate and encyclopaedic data about specific states—their size, shape, location, terrain, climatic resources, stage of development, government system, military forces, civic attitudes—acquires political

significance only with reference to some set of policy assumptions regarding the demands which they are likely to make on other states, and/or the demands which other states are likely to make on them. (1956, p. 49)

There seem to be three aspects of the study of government policy by political geographers. First there is the extent to which geographic factors are considered in making any policy decision. This is not only logically the first aspect, but it has also received most attention by political geographers. Second it is necessary to study the influence which geographical factors have on the operation of policy. This distinction was made clear by Sprout (1956, pp. 58–71, and 1957), who indicated that the policy-maker could evaluate only those geographical factors which were perceived. This leaves the possibility that certain factors which were not perceived could be significant when the policy was applied. This was certainly the case during the establishment of the British Groundnut Scheme in Tanganyika in 1948–9. The abrasive nature of Kongwa soil quickly destroyed ploughs and hoes, increasing costs, while its capacity to compact during the harvesting season impaired the efficiency of mechanical collection. Lastly the vagaries of the rainfall in the area had not been thoroughly established and sub-average years made the situation impossible. Third, political geographers must study the influence which the operations of policy have upon the cultural landscape.

If it is accepted that the study of policy should be developed along these lines the first question concerns the types of policies to be studied. It must be immediately agreed that political geographers have no right to pre-empt this field, which will be of increasing interest to economic geographers, as more and more governments assume greater powers in regulating the economic life of their countries. As Cohen suggests, however, economic geographers will probably be at least equally concerned to investigate the policies of private firms. This is a field which will generally lie outside political geography, the only exceptions will be the policies of very large companies with international operations, which may be significant in studying the political geography of specific countries. The significance of the policies of the major oil companies to the political geography of the Middle East is an obvious example of this point. But it seems likely that economic geographers will investigate the same aspects of policies as political geographers and both branches of the subject should benefit from the common approach and similar methods. Political geographers will be mainly concerned with the dual primary policies of independent states: the preservation of territorial integrity and the maximum development of the state's resources for the benefit of the population. Pounds (1963, p. *v*) refers to these aims as 'self-preservation and welfare'. The policies of national governments which do not impinge on these two aims are likely to be marginal to political geography. It is immediately apparent that both these aims will be served by policies operating within the state or outside the state, by policies which are normally described as domestic and external. The country is made stronger by the reduction of serious regional political differences within the state as well as by the conclusion of military alliances, or the establishment of overseas bases. Minority movements are often a source of weakness to states, especially if they occur in border areas close to unfriendly states. Consider the problems which the Kenya Government faces as a consequence of the political attitudes of Somali in the Northern Frontier District. Development is likewise pursued by policies which operate within the country and outside it, and these are so well known that they require no examples.

If this identification of dual aims is followed we will be able to break away from the traditional division into internal and external policies, which has been evident in many post-war studies (Moodie, 1947; Hartshorne, 1950). Moodie noted the problem of distinguishing between the external and internal political geographies of countries and justified it only on practical grounds, which would still apply if the political geographer wished to focus attention on only one aspect of the state. Millar (1967) expressed the difficulties which political scientists face in trying to identify all that is apprehended by the term 'foreign policy'. One reason for not discarding this traditional division in political geography and political science was advanced by Spykman (1942b, pp. 16–17), who explained that the important difference between the internal and external sphere of operations of any government lay in the order and authority at the national level, and the absence of overriding authority and of an established code of laws at the international level. This was a theme which Spykman frequently stressed (1939, 1942a) and which Moodie noted concerning the autonomy of states in respect of internal economic development. More recently Professor Greenwood in the 1966 Roy Milne Memorial lecture noted that internal policies were usually capable of open examination whereas foreign policies often had a shroud of secrecy imposed by the foreign governments concerned. But there is no suggestion that the order and authority of internal policies or their exposure to debate will necessarily result in a higher measure of predictability, or that the successful outcome of policy can be assured. There is a basic unity of national policy provided by the aim of creating the most favourable condition for the state. Any attempt to draw lines between domestic and foreign policies will face a number of difficulties. First, external geographical factors are often prime factors in determining domestic policies of development or defence. Second, domestic policies on questions such as tariffs or the treatment of minorities will influence the attitude of other countries. Third, some policies such as immigration will be difficult to classify as either domestic or foreign.

There is no conflict between these proposals and Hartshorne's suggestion that 'the fundamental purpose of any state . . . is to bring all the varied territorial parts into a single organised unit' (1950, p. 104). Hartshorne was concerned with the identification of the diverse regions of any state and this is still an important aspect of any analysis. But it is also important to follow this study with an examination of the policies adopted by the state to achieve unity, and to examine the effect of such policies on the continuing diversity of the regions and the cultural landscape. Hartshorne was obviously aware of these points, but he restricted his comments to 'the internal organisation of political authority . . . to permit different adaptations of government to different regional attitudes and interests' (1954, p. 199). There are other types of policies, including war and financial inducements, by which states will seek to overcome the political problems of regional diversity.

The second question concerns the way in which this material should be organised to simplify the co-ordination of research and to promote understanding. It is clear that there are three elements which may form a continuous chain: geographical factors, policies, and the geographical effects of policies. We can note that the new or altered geographical facts at the end of the chain may influence policy-makers in other countries, setting off a chain reaction, or provide the geographical factors to be taken into account by subsequent governments. An example of the chain reaction is provided by the maze of policy-decisions which followed the unilateral declaration of independence by Rhodesia.

Scholars concerned with the influence of geographical factors on policy will dissect out the influence of individual factors, such as size, location, and quality of population. This is the type of organisation which can be found in most texts of political geography. Students dealing with the effects of the operation of policies on the cultural landscape will arrange the material according to the aspects of the landscape affected: communications, distribution of population, industrial development. But both groups of research workers will have to start with one or a number of policies. Policies occupy the strategic position in the chain from which the backward view towards geographical factors and the forward view to the geographical influences of policy can be taken. McClosky made a similar point as a political scientist:

> decision-making . . . is usually a critical point in the process of international politics—a point of 'input' where the several influences that have gone into the decision can be detected and their relative effectiveness measured; and a point of 'output', where policies are unleashed and begin to register their effect on the course of international affairs.
> (Snyder, 1962, pp. 193–4)

It therefore seems essential to have some understanding of the nature of policy which will simplify the comparison of research results and the construction of a body of basic knowledge. This does not refer only to research within political geography, but would include research by economic geographers and political scientists. During the past decade many political scientists have focussed on decision-making as a central theme in their subject (see Snyder, 1962).

There are four qualities common to all policies—motive, method, subject and area of operation. From the point of the political geographer there seem to be three basic motives. Those of defence and development have already been mentioned; a third category includes policies of administration or organisation. They could theoretically be described as policies of development in the widest meaning of this term, but their distinctive nature makes their separation worth-while. This point was well made by Professor Spate in discussing a paper by Prescott (1967). Administrative policies will be concerned with the subdivision of the state into territorial units for purposes of local government, elections and the provision of basic services; it will also include policies connected with the choice of a capital, or an official language, or the division of powers between central and state governments in a federation. If a policy cannot be fitted into one of these three groups it is likely to be marginal to political geography.

Reference to method in this case is not made in the sense used by Spykman, who identified techniques of coercion or negotiation. A more meaningful division for geographers is between unilateral decisions made by a single government, and multilateral decisions reached after consultation or involvement of more than one government. The significance lies in knowing whether it is necessary to construct one geographical view or more than one, for different governments will view the same geographical facts in different ways. The importance of this position was made clear in an exasperated letter written by Balfour to Lockhart in June 1918.

> You constantly complain of indecision, as if all that was required was that H.M.G. should make up their minds. But there has been no particular indecision on the part of particular members of the Alliance. They have severally determined their policy as quickly as could reasonably be expected . . . Britain, France and Italy have thought the dangers of intervention less than its advantages; America has thought the advantages less than the dangers; Japan will do nothing on the grand scale until she receives an invitation from her co-belligerents.
>
> (Quoted in Ullman, 1961, p. 192)

This division between unilateral and multilateral methods might seem to parallel the division into domestic and foreign policies. However, it is clear that governments may take unilateral action outside their own territory if they are sufficiently powerful, and conversely many governments have engaged in multilateral agreements relating to their internal development.

There is a very wide range of policy subjects, such as trade, conscription and investment control, and it is impossible to give an exhaustive list.

The area of operation of a policy includes both the geographical area and the section of the economy or population to which the policy applies. For example, policies regarding 'beef roads' in the Northern Territory of Australia apply to a clearly defined area of land, while restrictions on the production of margarine apply to a particular section of the country's manufacturing industry wherever it may be located, and conscription applies to a particular section of the community wherever they may live. In addition to the area of policy the geographer must also know its duration.

Four qualifying points must be made to this suggested view of the qualities of policies. First, some policies will serve more than one motive. South Africa's search for oil has the defensive aim of strengthening the country's resistance to any economic sanctions, and the economic aim of improving the balance of payments by reducing the level of imports. Second, while the subjects of a number of policies in different countries may be the same they may serve different motives. If control over industrial location is examined, it is clear that in Britain this policy is designed to help the economic stability of some depressed areas and reduce the problems of planning in the Midlands and south-east England. In South Africa the establishment of border industries around the Bantustans is designed to reduce African concentrations around the main cities and reduce international criticism of apartheid policies. In Germany between the two world wars autobahns were built for defence and improved military efficiency, improvements in Irish roads are designed to make travel easier for tourists. Third, there will not necessarily be any correlation between motive and effect. For example, defence policies may have economic consequences. Britain has indicated that its timetable of defence cuts in Singapore took account of the economic significance of the base to the revenue of the island. Fourth, it is not necessary for geography to have been significant in the formulation of policies for their implementation to have geographic consequences. The doctrinaire nationalisation of industry will frequently produce geographical consequences, and any rise in the price of gold, that is based on political and financial reasons, would have a profound effect on the pattern of gold extraction. If it is accepted that Indonesia's policy of confrontation against Malaysia was compounded mainly of political and ideological elements, it is important to note that the effects of the policy included the cessation or hindrance of trade between the two countries, the construction of new roads and airfields in Sabah and Sarawak, and the evacuation of the Malaysian border zone accompanied by the regrouping of the Chinese population in supervised settlement areas.

The third question concerns the hypothesis to be used in establishing the relation between geographical factors and policy decisions. Sprout (1956) has made a detailed analysis of the man-milieu approaches available in respect of international politics which has value for political geography, because international politics includes the calculation of state power and the explanation and prediction of state actions. There is little doubt that examination of cognitive behaviour offers the best opportunity, providing the individual decision-makers can be identified, and providing there is sufficient material on which to reconstruct their appreciation of the significance of geographical factors. This will rarely be possible except in historical cases when material from archives is available, therefore it will usually be more fruitful to employ a concept of probabilism which utilises a general model and makes assumptions about motives, skills and knowledge. The difficulty of identifying the particular decision-maker in some cases may be gauged from the very complex diagram published in *The Times* (14 November 1966) to illustrate the economic corridors of power in Britain. Political scientists have often written on the problems associated with the meaningful analysis of policy decisions.

> The matter becomes more complicated from here on, for many different variables have to be taken into account in assessing the influences on any particular decision. Psychological, social and economic factors may need to be investigated, in addition to the usual political ones.
> (McClosky, writing in Snyder, 1962, p. 194)

> The difficulties of getting at the facts in foreign policy before the files are opened are obvious enough. Australia is one of the hardest of the democracies in this respect . . . There are no immutable or absolute factors in foreign policy. This is what makes writing about foreign policy so difficult. Perhaps nations ought to determine their policies in accordance with set principles—geographic, demographic, military, economic, ideological and so on. Prime Ministers ought to be rational, however that may be judged, but they are not always so. To find the basis for the foreign policy of a country, therefore, it is necessary to ascertain why relevant decisions were actually made. This means looking at the thinking of people who made the decisions, their image of the world and their own policy, of finding which facts were factors to them, and how they took them into account.
> (Millar, 1967, pp. 73–5)

Some political scientists have used the technique of circulating manuscripts dealing with contemporary events, amongst politicans involved, inviting comment. Gross errors may be avoided in this way, although it is necessary to beware of politicians trying to present their policies in the best possible light.

It seems worth while to examine in more detail the nature and value of historical and contemporary studies. While it will be much harder to make correct assessments of the significance of various factors in any contemporary policy decision, than in the case of decisions made so long ago that the archives may be consulted, the information about the details of the policy and many of its effects will be equally available in historical and contemporary studies. Sawer (1967) has made the point that

> there is no reason whatever to doubt that the ultimate expression of policy is reliable, and that the behaviour of many governments will be in accordance with the policy so expressed.
>
> (p. 236)

In terms of the effects of policies, the only difference between historical and contemporary studies will be that in the former it will be possible to include long-range influences, and provide a fuller assessment of the extent to which the policy results were predictable.

In any case, political geographers have a duty to consider the contemporary scene, despite the attendant problems. This point was stressed by Moodie:

> . . . the political geographer is concerned with the observation, recording and analysis of the changes in the world which have already taken place, as well as those which are proceeding at the present time.
>
> (1947, p. 12)

It is recognised that contemporary explanations might be controverted by later scholars using primary sources, but such later scholars will be grateful for the impressions recorded by contemporary workers, and the descriptions of policy effects should be accurate. These contemporary studies will also assist in the examination of the relationships between policy and geography at different points in time within a single state. It would be interesting, for example, to test the views of the German statesman Kuhlman:

> . . . the geographical position and historical development are so largely determining factors in foreign policy, that regardless of changes in the form of government, the foreign policy has a natural tendency to return again and again to the same general and fundamental alignment.
>
> (Quoted in Sprout, 1945, p. 63)

It is presumably this belief which accounts for the persistence of Mackinder's Heartland theory, with modifications, since 1904. A further reason for conducting contemporary research is that it will provide means of assessing the significance of changing technical skills.

The main contemporary sources will be government reports, including proceedings of parliament, collections of documents and treaties, the memoirs of politicians and newspapers and journals. These sources must be continually cross-checked and it is essential that the student is meticulous in providing references for the facts and views used, so that other students may make maximum use of the study and assess its validity. If this is not done it will be very difficult to co-ordinate research amongst political geographers, and between political geography and other disciplines, such as economic geography and political science. The absence of contemporary accounts in the past has hindered research into historical studies in political geography. It may not be fashionable at the present time merely to describe the main features of a country's political geography, such as its administrative division, electoral patterns, trade agreements and defence alliances, but this is as much part of political geography as analysis, and very difficult to do well.

The fourth question concerns the extent to which the political geographer may search for laws or general principles regarding the relationships between geography and policy. In the answer to this question is found a major difference in the approach to policies of political and economic geographers. Economic geographers have such a large number of firms to consider that they can confidently expect to establish certain well-based principles. In the case of political geography there is the problem that while there are a large number of political authorities they tend to exist at clearly defined levels. The vastly different levels of responsibility between national and local governments probably prevents any useful generalisations about the relationships between geography and policy. For example local authorities, including state governments in a federation, do not usually have any responsibility for state defence. This question has been the subject of debate amongst political geographers. Goblet (1955, p. 20) has urged geographers to seek laws affecting the rise and fall of states, while Hartshorne (1950, p. 128) maintains that political geography lacks the multiplicity of examples to allow the establishment of cause and effect relationships.

There is insufficient evidence at the moment to make any final pronouncement about the extent to which studies of local and national government are comparable.

Some political scientists, for whom Wright (1955) is an eloquent spokesman, have no doubt that political geography is unable to produce any 'conceptual systems'.

> It has been the hope of some geographers that because of the apparent permanence of geographic conditions, geography might become the master science of international relations. This hope seems vain. Geography is primarily a descriptive discipline. It provides evidence, as does history, for generalisation by the disciplines of politics, economics, sociology, and psychology, and it may itself make a few empirical generalisations. Geography, however, does not determine international relations. . . . Geography cannot develop concepts and conceptual systems applicable beyond a limited time and area in which a given state of the arts, of population and of society can be assumed.
>
> (p. 348)

In view of these strictures against political geography, it is surprising that Wright, who regards the sociology of international relations as being 'most successful in establishing theoretical bases for a science of international relations' (p. 334), should praise the sociologists in the following terms:

> Sociologists do not assume the inevitability of any institution nor of a *human nature* which is implacably aggressive, implacably irrational, or implacably incorrigible. Institutions are the consequence of historical contingencies and they change as men plan and history proceeds. Human behaviour, while based upon the biological character of man, is a consequence, so far as international relations are concerned, of technological, cultural and institutional conditioning.
>
> (p. 399)

These quotations demonstrate that political scientists in many cases have a poor impression of political geography; this point will be raised later.

The fifth question concerns the extent to which the geographer should concern himself with future policies. Such a question immediately raises thoughts of *Geopolitik* which will cause many geographers to shy away, but the question cannot be baulked. Writing about international politics, Sprout (1956) expressed the following view.

> Neither the statesman, nor the government research analyst nor the academic student of international politics can escape the future. Every opinion, every proposal for action, every decision [on] foreign policy reflects some image of the future, some expectation of how proposed solutions might affect the shape of things to come.
>
> (p. 87)

This statement also applies to geographers. If it is admitted that in certain policy decisions it is important that geographical factors should be considered, and that policy operations will have influences on the environment and landscape, then there is no one better qualified than the geographer to present these factors and evaluate these effects. Goblet and Hartshorne have both commented on this aspect.

> On the one hand it [political geography] makes a critical geographical analysis of existing States and of those in course of transformation or about to be transformed, and studies the diplomatic actions designed to recast the political map of the world. On the other hand, it prepares schemes for the delimitation of the optimum territorial extent of each state, bearing in mind its geographical environment and its relations with others. In this way it may be regarded as an 'information service' for governments and their diplomatic corps. . . . But it is outside the scope of political geography to put forward solutions. . . .
>
> (Goblet, 1955, p. 20)

> If plans are being made for the construction of an entirely new state-area, or for major territorial alterations in an existing one, one is forced to attempt some prediction of the capacity of such a projected organisation to function effectively as a unit. Political geographers will be able to claim superior competence in attempting predictions in such cases only if they have established a high degree of understanding of the reasons why present or past state-areas have or have not functioned effectively.
>
> (Hartshorne, 1950, p. 129)

Disregarding Goblet's second aim, which seems questionable, these two quotations suggest the guide lines for geographers engaged in studying geographical factors relating to future policy decisions and their operation. First, the geographer's special knowledge and techniques do not confer any special ability to specify policy aims or recommend the way in which those aims should be achieved. Clearly policy aims and methods will be based on both geographical and non-geographical factors just as national power includes non-geographical factors such as morale and the quality of leadership, a point which East (1950) felt Mackinder neglected. All the geographer can legitimately do is to say in respect of any particular policy problem, these geographical factors seem to be significent to me, and these geographical consequences are likely to follow these decisions. In this sense Goblet is right in saying that geographers will act as an intelligence service, in precisely the same way as economists and political scientists. This view agrees with Spykman who wrote:

> . . . the adherents [of *Geopolitik*] are not only engaged in a study of the geographic conditioning of political phenomena; they are also engaged in advocating policy, which is hardly a scientific endeavour.
> (1938, p. 30)

This was a view which Spykman abandoned in some of his later studies of American strategy. The predictions to which Hartshorne refers are probably similar to those which were described by Sprout (1956) as non-policy oriented predictions. In such cases a set of existing conditions (or presumably, proposed conditions) are projected into the future and their effects deduced. Wise (1965) demonstrated the validity and usefulness of this technique in his examination of the geographic implications of a tunnel under the English Channel, to the planning of development in south-east England. Such an exercise contrasts with policy-oriented decisions, where the student begins with the desired condition to be achieved, and estimates the changes, environmental and others, which are necessary to achieve it. This last is clearly an examination of method to which the geographer's contribution must be limited to a presentation of the geographic facts which seem to be important. Any predictions which are made should be compared with subsequent reality so that the techniques for accurate prediction can be improved. Sprout (1956, p. 90) has also drawn attention to an important difference between analyses of historical and future situations. The analysis of historical examples cannot change what happened, it can only throw more or less light on the relationships between influences and results, but the future can be changed, and it is possible that convincing presentation of geographic facts may encourage the adoption of one policy rather than another. Although research into historical and contemporary situations should be equally objective, it is salutary to bear the possible significance in mind.

It is also worth reiterating that the political geographer cannot allow himself to make moral judgements on policies, the need for objectivity here, as throughout the subject, is fundamental. This interpretation of the attitude of political geographers to future policy-making finds support amongst scholars studying other branches of the subject.

Meyer (1963) published an article on the contribution of urban geography to planning. It is clear that Meyer is primarily concerned with the geographer's role in providing information through concepts of potential and hierarchy. For example, he refers to isoline and isochrone maps, to studies of central business districts, and to the importance of range and threshold concepts to the planning of internal urban transport. Although he concludes by claiming that the 'geographer is a fully fledged member of the planning team', he does not outline any claims to participation in the making of policy. Presumably the geographer's influence on policy-making will be through forceful presentation of painstaking research programmes.

Bird (1962) discussed the utilisation of some Australian lakes, and showed how there was often a conflict between the needs of various activities such as tourism, conservation, agriculture and industry. He did not make any firm policy recommendations, but usefully indicated the way in which past developments have influenced, and future developments may influence, the resource value of lakes. Forrest and Drew (1966) initiated a geographical analysis of the planning concept contained in certain planning acts and regulations, that the best soils in New Zealand should be preserved for agricultural use. They explored the differential costs of establishing new suburbs on coastal hills rather than the fertile Taieri Plain north-west of Dunedin. The only firm recommendation by these authors is that a national survey and full economic analysis of this problem is necessary before the final formulation of government policy.

In his consideration of geography and planning, Freeman (1958) is concerned to show how a study of geography is vital to effective planning and administration. He cogently demonstrates the geographical factors which are most important, and indicates the conflict which may develop between various uses of land. But there is no claim that the geographer should play any role in deciding what the aims of policy ought to be, or settling how any policies should be implemented.

An example of how geographers have occasionally been more

directly concerned with policy is provided by studies on the administrative boundaries of the United Kingdom. Fawcett (1917 and 1919), Gilbert (1939) and Taylor (1942) not only advocated the reorganisation of administrative boundaries because of their failure to meet contemporary needs, but clearly laid down the principles by which such boundaries should be selected, although they must have been disappointed with the regard given to their recommendations by the authorities in subsequent boundary changes.

The views of Ackerman (1962) conveniently summarise this brief discussion. In a useful survey of geographers' contributions to the formulation of public policy, he reaches the conclusion that involvement is a matter of personal conscience and predilection. He then stresses that geographers have made helpful contributions since 1919, that geographers may continue to do so, and that in cases where the issue is vitally related to geography, geographers have a duty to make their conclusions known. This would certainly include the right to criticise a policy, before it was implemented, if it was thought that it was out of balance with the geographical facts, if, for example, it was thought the policy was being launched from too narrow a resource base.

Having advocated that political geographers should show a greater awareness of the individual policies of individual governments, it is now necessary to outline the advantages which would follow such a development. First, it would enable political geographers to conduct more detailed research and avoid some of the fallacies which characterise general statements about geography and state policy. Second, it would give an increased measure of realism to geographical research because it would consider concrete issues, and at the same time stimulate the interest of non-geographers. Third, it would assist co-operation between political, economic and urban geographers. Fourth, it will facilitate co-operation between political geographers and political scientists. This seems to be a most important point, for examination of a selection of texts by political scientists interested in international politics reveals a number of disquieting features.

Political scientists tend to take a very narrow definition of geography.

> The second element of national power is the geographic: where do people live and what climate do they have, how have they been influenced by their geographic location (and in turn affected their

geographical environment), and what is the size of their territory?
(Hartmann, 1957, p. 47)

While Morgenthau (1966, pp. 110–12) records that geography is the most stable factor upon which the power of a state depends, he only considers size, location and the existence or absence of 'natural frontiers'. Wright (1955), whose views have been quoted earlier, devotes fifteen pages to a statement of political geography, which is defined as 'the science relating the physical environment to politics' (p. 336). This very old-fashioned definition is followed by a number of statements about the aims and methods of political geographers, with which most political geographers would disagree.

As a political geographer usually writes from the point of view of his own state, he often seeks to explain or criticise past policies and advise on future policies of his state from an analysis of its geographical position, and to predict the probable policy of other states, particularly political rivals, from an analysis of their geographical position.
(p. 338)

Most political geographers have, however, adopted a national point of view, partly because as citizens they have put their own state's interests first, partly because as historians they are more familiar with national than international politics, and partly because as scientists they have interpreted their function as prediction rather than control....
(p. 338)

In common usage geography is a description of the earth's surface with special reference to those features of human or social importance. In this sense it is an aspect of history, for the latter deals not only with the sequence of events in time, but with their location in space, and with their explanation not only by temporal antecedents but by spatial proximities. Apart from the techniques of mapping to manifest spatial relations, geography in this sense may be considered history which emphasises the influence upon human behaviour and institutions of the physical environment and spatial relations. Political geography from this point of view consists of the interpretation of political boundaries and of political history in terms of location, situation, and environment.
(p. 336)

Morgenthau (1966) and Hartmann (1957) consider that the distribution of people and resources lie outside geography, but in their treatment of these aspects they both neglect the significance of national quality or degree of resource development, features which geographers have studied with success.

The second striking point is the way in which the importance of geography may be stressed at the beginning of the text by the political scientists, then neglected during the case studies which are included. Lastly, it is clear that the geographers who have attracted most attention from political scientists are those concerned with global strategies, such as Mackinder (1904 and 1919) and Fairgrieve (1941), and geopolitics, such as Weigert (1942). Some of the blame for this misinterpretation must attach to the political scientists, but some remains with political geographers, who have not pointed out these errors, and who have failed to demonstrate more clearly that they can undertake objective research into the relationships between geography and politics, of which government policies form an important part.

REFERENCES

ACKERMAN, E. A., 1962, 'Public policy issues for the political geographer', *Annals*, Association of American Geographers, **52**, pp. 292–8.

BIRD, E. C. F., 1962, 'The utilisation of some Australian lakes', *Australian Geographer*, **6**, pp. 199–206.

COHEN, S. B., 1966, 'Toward a geography of policy', *Economic Geography*, **42**, Guest editorial for January.

EAST, W. G., 1950, 'How strong is the Heartland?', *Foreign Affairs*, **29**, pp. 78–93.

FAIRGRIEVE, J., 1941, *Geography and world power*, 8th ed., London.

FAWCETT, C. B., 1917, 'Natural divisions of England', *Geogr. J.*, **49**, pp. 124–41.

FAWCETT, C. B., 1919, revised ed. 1961, East, W. G. and Wooldridge, S.W. (Eds.), *Provinces of England*, London.

FORREST, J., and Drew, E. A., 1966, 'Urban spread: the economic use of New Zealand's first class soils', *Australian Planning Institute J.*, **4**, pp. 16–19.

FREEMAN, T. W., 1958, *Geography and Planning*, London.

FREEMAN, T. W., 1968, *Geography and Regional Administration*, London.

GILBERT, E. W., 1939, 'Practical regionalism in England and Wales', *Geogr. J.*, **94**, pp. 24–44.

GOBLET, Y. M., 1955, *Political geography and the world map*, London.

HARTMAN, F. H., 1957, *The relations of nations*, New York.

HARTSHORNE, R., 1950, 'The functional approach in political geography', *Annals*, The Association of American Geographers, **40**, pp. 190–201.

HARTSHORNE, R., 1954, 'Political geography', in *American geography, inventory and prospect*, James, P. E. and Jones, C. F., Syracuse.

JONES, S. B., 1954, 'The power inventory and national strategy', *World Politics*, 6, pp. 421-52.

MACKINDER, Sir H. J., 1904, 'The geographical pivot of history', *Geogr. J.*, **23**, pp. 421–44.

MACKINDER, Sir H. J., 1919, *Democratic ideals and reality: a study in the politics of reconstruction*, New York.

MEYER, H. M., 1963, 'Urban geography and urban transportation planning', *Traffic Quart.*, pp. 610–31.

MILLAR, T. B., 1967, 'On writing about foreign policy', *Australian Outlook*, **21**, pp. 71–84.

MOODIE, A. E., 1947, *Geography behind politics*, London.

MORGENTHAU, H. J., 1966, *Politics among nations*, New York.

POUNDS, N. J. G., 1963, *Political geography*, New York.

PRESCOTT, J. R. V., 1967, 'The reciprocal relations between geography and national policy', paper delivered to Section P. (Geography), *Australian and New Zealand Association for the Advancement of Science*, Melbourne.

SAWER, G., 1967, 'On writing about foreign policy: Comments', *Australian Outlook*, **21**, pp. 235–7.

SNYDER, R. C., Bruck, H. W., Sapin, B., 1962, *Foreign policy decision-making*, New York.

SPROUT, H. H. and M., 1945, *Foundations of national power*, New York.

SPROUT, H. H. and M., 1956, *Man-milieu relationship hypotheses in the context of international politics*, Princeton.

SPYKMAN, N. J., and Rollins, A. A., 1939, 'Geographic objectives in foreign policy', *Amer. Pol. Sc. Rev.*, **33**, pp. 391–410 and 591–614.

SPYKMAN, N. J., 1942a, 'Frontiers, security and international organisation', *Geogr. Rev.*, **32**, pp. 436–47.

SPYKMAN, N. J., 1942b, *America's strategy in world politics*, New York.

TAYLOR, E. G. R., 1942, 'Discussion of the geographical aspects of regional planning', *Geogr. J.*, **99**, pp. 61–80.

ULLMAN, R. H., 1961, *Anglo-Soviet relations 1917–21: intervention and the war*, Princeton.

VAN VALKENBORS, S., 1939, *Elements of political geography*, New Jersey.

WEIGERT, H. W., 1942, *Generals and geographers*, New York.

WISE, M. J., 1965, 'The impact of a channel tunnel on the planning of S.E. England', *Geogr. J.*, **131**, pp. 167–85.

WRIGHT, Q., 1955, *The study of international relations*, New York.

2

GLOBAL POLICIES

One field of political geography which has attracted considerable attention, and which is directly related to state security, is global strategy. Academics who have presented analyses of global strategy have generally done so with the intention of influencing policy. This has either been done by drawing attention to areas which seem to offer a potential threat, and leaving the policy-makers to draw the final conclusion, or by specifically advocating policy. For example, Mackinder (1919, p. 2) was concerned that the facts of geography encouraged the growth of world empire, and that the influence of such geographic factors must be countered by the action of interested states. In a recent review East (1965) suggests that geographers have a continuing responsibility to contribute to the assessment of the changing strategical importance of various parts of the world. Geographers are well equipped to make a major contribution, providing two qualifications are remembered. First, each state takes an egocentric view of the world, and therefore there are as many global views as there are states. Second, since global views are meaningless unless they are associated with an assessment of the relative power of states, geographers cannot expect to provide the final answer alone.

The analysis of national power represents a distinct area of convergence, not only of geography and political science but also of economics, anthropology and psychology.

(Hartshorne, 1954, pp. 175–6)

The importance of Mackinder's work, in the field of global strategy as a bench-mark and as a source of inspiration to others, whether geographers or students of international relations, is well known. All his writings have been carefully analysed by a number of very competent geographers (Weigert, 1949; East 1950), and therefore this material will not be considered again in detail. Instead it is proposed to focus attention on three more recent studies, which all owe something to Mackinder's pioneer work. In order to emphasise that this is a field where many subjects converge only one of the studies will be by a geographer. The studies are by Spykman a political scientist, Mouzon an economist, and Cohen a geographer. The major studies of each were published after the beginning of the Second World War, and each took an American-centred view of global strategy.

Spykman published four papers and two books dealing with foreign policy and geography, in the decade before his death in 1943. The earliest paper (1933) is quite distinct from the other material, which appeared after 1938, in considering the methodological foundation of the study of international relations. Convinced that the study of international relations is a social science, Spykman outlines an approach which satisfies his criteria of scientific enquiry: is a correct inference possible from observable data, and can the inference be expressed as a concept with general validity? His view is that the core of the subject is the political behaviour of states, as manifested in their foreign policies. Assuming that foreign policies are a function of international (external) and national (internal) factors, Spykman advocates the measurement of the importance of varying international factors, by an analysis of the state's foreign policy when the internal factors remain constant. By the same method, it would be possible to measure the importance of varying internal factors, during a period of unchanged external situations. The comparison of these assessments for similar states would permit a 'statistical prediction' or indicate a 'regularity of correlation' of state political behaviour, and assist in understanding why states behave as they do. Clearly no law could be formulated due to the impossibility of isolating single factors in either the external or internal environment. Morgenthau and Thompson (1956, p. 24) have praised Spykman as a pioneer, in placing international politics at the centre of the study of international relations, and note that subsequent trends have confirmed his view. This contribution to the definition of the subject does not appear to be matched by the approach suggested. It seems questionable whether internal or external conditions, ever remain constant for a sufficiently long period to allow assessment of various groups of factors, except in the case of a very few states, which are not important in world affairs. The reader may care to consider, to which periods in the history of the states he knows best, the research programme outlined by Spykman could be applied. Even if conditions did allow this analysis for some states, it is likely that the uniqueness of all states would prevent the abstraction of other than highly generalised concepts. In his own subsequent studies, Spykman restricted his analysis to the correlation between the geography of the state and the past or proposed foreign policies.

Three papers by Spykman published after 1938 considered 'Geography and foreign policy' (1938), 'Geographic objectives in foreign policy' (with Rollins, 1939) and 'Frontiers, security and international organisation' (1942a). It is proposed to summarise the three papers and then make a composite valuation. In the first of these papers, Spykman reaffirms his views on the best methods of research into international relations.

> It is the task of the social scientist to try to find in the enormous mass of historical material correlations between conditioning factors and types of foreign policy. This means that the study of diplomatic history must be supplemented by a search for the behaviour patterns of states under different stimuli and in various international environments. Scientific method requires that the search operates by means of abstraction, and common sense warns that correlations found by means of abstractions can by themselves be only partial, not complete, explanations of concrete historical situations.
>
> (1938, p. 28)

He then sets himself to examine the correlation of foreign policy with geography which he regards as the most fundamentally conditioning factor because of its relative permanence. In this matter he follows Mackinder, who recognised that the balance of power in the world was a function of geography, people and political organisation, but held that 'the geographical quantities in the calculation are more measurable and more nearly constant than the human' (Mackinder, 1904, p. 437). Geographic factors are then considered under the headings of size and location, the latter being subdivided into world location and regional location. World location considers the territory in respect of major climatic zones, seas and land masses; the regional location in relation to neighbouring states. These very broad divisions include brief treatment of topics which would have been worth more detailed, separate consideration. Thus territorial shape is considered in one paragraph, that refers only to the ideal and most disadvantageous forms. Most of the section considering the factor of size deals with the problems of unifying large areas, which is largely, though not exclusively, a matter of internal policy.

The factor of location is treated at greater length and more systematically. Spykman suggests that in respect of location, it is necessary to determine both the facts and their significance, during the period under consideration. The facts of location do not change but the significance of location can be altered, by new discoveries and technological advances. In examining the world context of states, Spykman concentrates on oceanic trade routes and climate, at the expense of topography and mineral deposits. States are classified into three types, each of which may have one of three basic regional locations. The three types are land-locked, island and marginal states, and the regional location depends on whether the neighbouring states are stronger, weaker or equivalent in strength. The various combinations are considered in differing detail, marginal states naturally receiving most attention. Topography receives more attention than before, in a concluding section, which also deals with the nature of borderlands—zones of exposure or protection—and their influence on foreign policy.

Spykman introduces his analysis of the geographic objectives of foreign policy (with Rollins, 1939), by referring to the inability of the League of Nations to provide territorial security for member states, and the desire of those states for certain geographic objectives. The nature of expansion is examined and boundaries are regarded as the 'political geographical expression of the existing balance of [state] forces' at any period. Shifts in boundary position, as well as changes in foreign policy and the granting of leases, are indications of changes in the balance of political forces. For the purposes of this paper only objectives requiring a change in boundary position are considered. A long section dealing with the nature of boundaries and frontiers shows how the former have developed within the latter, and many examples are provided. Then follows three types of boundary extension for particular motives—to include an entire river basin within the state's area, to secure access to the sea, and to control opposite or adjacent coastlines.

Spykman's paper entitled 'Frontiers, security and international organisation' was read to a joint session of the Association of American Geographers and the American Political Science Association, and published in the *Geographical Review* (1942a). After reiterating that the independent state is responsible for its own territorial integrity, and that the boundaries of any state are an indication of the power differentials of adjacent states, Spykman reviews the three main methods of neutralising power differentials —by individual action, by alliances and by systems of collective security. Each of these systems is considered in turn, and it is shown how technological advances in weapons have reduced the value of relief and buffer states as first lines of defence, and how tested systems of collective security have failed, when there are considerable power differentials between the attacking and the threatened state. In conclusion, Spykman suggests that greater stability would be ensured if it was possible to create states of equivalent power potential. However, he recognises that the difficulties of drawing boundaries to provide this situation, make the solution impractical, and he calls for co-operation between geographers and political scientists to formulate the best working solution for a peace that is recognised to be impermanent.

It is clear that in these three papers there has been a steady change in Spykman's interest from the simple study of international relations which he outlined in 1933. He applied his methodology for international relations in his first study (1938), but considered only the relatively unchangeable and measurable geographical factors. In the second paper he was concerned with the geographic aims of states, and lastly he considered the power potentials of states and outlined some methods by which maximum security for most could be achieved. Spykman's cultivation of the field linking foreign relations and political science on the one hand, with geography on the other, means that his work must be assessed in the light of political geography as it existed then. Such an assessment indicates three points.

First, most of the facts presented in Spykman's papers, and the systems under which they were organised, were not fresh to the content of political geography. Spykman referred mainly to the works of Hennig (1931), Bruhnes and Vallaux (1921) and Maull (1925), ignoring the important studies of Bowman (1921), East (1937), Hartshorne (1935), Ancel (1938) and Ratzel (1895). It is this last omission which is most surprising. To write a paper which considers the size and location of states, without referring to Ratzel's exhaustive treatment of *der Raum* and *die Lage*, is to display surprising ignorance or indifference to the standard methods of presenting research results. Many of the ideas expressed in these three papers and in Spykman's later books can be found in literature published earlier, yet often no acknowledgement was made, a failure which provoked criticism from Earle (1943) and Sprout (1942) when they reviewed Spykman's major work, *America's strategy in world politics.* Spykman may have reached his conclusions independently, but in that case he had done very little reading in his subject. This is just a possibility, for the second point is that Spykman made some errors, which should have been avoided by persons familiar with the literature of political geography. In examining the treatment of boundaries it is surprising to note that there is a failure to distinguish between frontiers and boundaries. While some political geographers were still using the terms as synonyms Lapradelle (1928), East (1937) and Hartshorne (1936) had clearly indicated the need for careful use of these terms. Further, in advancing his concepts of boundaries, as the expression of the existing balance of forces, Spykman was making the same error as Ancel (1938), who described boundaries as *isobares politiques.* It is not true to write in the following terms:

> Boundary changes will be indications of a shift in the balance of forces, caused either by an increase in driving force on one side of the frontier [boundary] or by a decrease in resistance on the other.
>
> (Spykman and Rollins, 1939, p. 392)

Lapradelle (1928) and others had shown that boundaries change position during the process of evolution through the states of allocation, delimitation and demarcation, without any change in the strength of the separated states. Boundaries are changed to provide simpler marking and for the mutual convenience of states. Examples are provided by the changes in boundary position in Africa, after the First World War. Britain and France varied the boundaries dividing the former German colonies of Togoland and Kamerun, because the initial division proved inconvenient. On the other side of the continent Britain freely gave the Juba strip to Italy, after World War I, as a reward for participation. A second error is found in Spykman's comments on mountains, as the frontier within which a boundary might be drawn.

> Mountains . . . provide a barrier zone in which it is comparatively easy to draw as the boundary the line of the peaks or that of the watershed. These two lines will not necessarily coincide, but their divergence will create a problem of only minor adjustment. Such a boundary has an economic as well as a strategic justification, for there is always an unmistakable orientation of a population towards a valley which they can see from the mountain crest.
>
> (Spykman, 1939, pp. 401–2)

Was Spykman not aware of the *cause célèbre* between Argentine and Chile over the interpretation of their common boundary in the Andes? This is a classical illustration of the simple fact that the mountain *crest* and the *watershed* may differ in position by tens of miles and lead to bitter controversies (Hinks, 1921). Further, it is naive to regard mountain crests as invariable economic divides. Frequently, boundaries drawn through mountainous areas dislocated the economic and social life of cultural groups occupying the area. Ryder (1925) noticed this in the Iranian borderland, and a more recent example has come to light in the Himalayan region, as a result of the Sino-Indian dispute (Kirk, 1962).

Some of Spykman's abstracted concepts do not seem to have general validity, as the following example shows.

> A peninsular position is conducive to the development of sea power, particularly if the land frontier of the state is protected by a natural barrier like the Himalayas, the Pyrenees or the Alps.
>
> (Spykman, 1938, p. 223)

The development of Indian sea power is very slight, and neither Spain nor Italy has a major fleet today, although it is conceded that in an earlier period Spain was a first-rank sea power. However, whether this was mainly due to its form and the protection offered by the Pyrenees is debatable.

The third point which arises concerns Spykman's discussion of *Geopolitik*, which at that time was attracting considerable attention in America. There are two references to *Geopolitik* in the first two papers, and both are derogatory.

> The geographical determination which explains by geography all things from the fourth symphony to the fourth dimension paints as distorted a picture as does an explanation of policy with no reference to geography. . . . The present German school of 'Geopolitik' has

abandoned to a certain degree the strict geographic determinism of Ratzel, but only to be tempted by a metaphysics which views geography as a last chance. As the word indicates, the adherents are not only engaged in a study of the geographic conditioning of political phenomena; they are also engaged in advocating policy, which is hardly a scientific endeavour.

(Spykman, 1938, p. 30)

The 'metaphysics' of the 'natural' frontier, notwithstanding its pretended scientific basis, has been more confusing than helpful . . . The very human habit of calling natural what is desirable, and viewing as unnatural that which is undesirable, still characterises many students of political geography. Moreover, the prophets who are leading their peoples in a quest for the 'natural frontier' are apt to forget that most frontiers have two sides.

(Spykman, 1939, pp. 398–9)

The second statement presumably refers to students of *Geopolitik* since several political geographers had convincingly attacked the concept of natural frontiers (Sölch, 1924; Maull, 1925; and Hartshorne, 1936) and Hartshorne (1935) had pointed to the dangers of subjective studies in political geography. Spykman (1938, p. 30) expressed his position in the matter, as lying between Ratzel's determinism and the humanist view of the French possibilist school, which believed that human will was the force which played the greatest role in determining man's destiny (Febvre, 1922, p. 51).

Geography does not determine, but it does condition, it not only offers possibilities for use, it demands that they be used; man's only freedom lies in his capacity to use well or ill or to modify for better or worse these possibilities.

(Spykman, 1938, p. 30)

This expression seems to lie very close to the French position. The freedom, which is acknowledged as belonging to man, is sufficient to make it a prime factor, and it is difficult to interpret the phrase 'Geography . . . demands that they be used' as meaning anything else than, every environment offers a limited range of opportunities for economic state development. Man's ability to transform the environment is considerable, but spectacular changes are often costly and would overstrain the economy of the state concerned. As Spykman's two books are reviewed, it will be seen that he apparently moves away from the position outlined above, and closer to the German school of *Geopolitik*. First, it will be noticed that geography is given first place and that the human factor is neglected. Second, it is apparent that Spykman begins to advocate policies, an undertaking which he did not at first regard as a legitimate extension of research in social sciences.

Spykman's major work, *America's strategy in world politics*, was published in 1942. The book is a geopolitical analysis of the basic dichotomy of American foreign policy—isolationism and intervention—in terms of geography and political power. In the first part of the book, the relations between America and the other centres of regional power, in the trans-Pacific and trans-Atlantic zones, are examined, and the main conclusions are then incorporated in a chapter dealing with the world context of American foreign policy. The second and longer part deals with the struggle for influence in South America, between America and other states.

Part one is based on four principles. First, the international environment is anarchical and states are responsible for self-preservation, which means the maintenance of territorial integrity (Spykman, 1942b, pp. 16–17). Second, the foreign policy which provides maximum security should be rooted in the geography of the country (Spykman, 1942b, p. 41). Third, the world consists of a closed political system and events in one area will have repercussions throughout the world (Spykman, 1942b, p. 165). Fourth, America must prevent the unification, through conquest, of the trans-Pacific and trans-Atlantic areas, in order to avoid encirclement and defeat. The detailed analysis, which is based on these principles, produces a policy of intervention and convincingly destroys any concept of isolationism. It is proposed that regional Leagues of Nations be developed in Asia, Europe and America, and that as far as possible states of equal power should be created, although it is noted that only in Europe will there be even a slight chance of doing this. In each case there will be an extra-territorial member—America in the case of the Asian and European systems.

The second part reviews the ideological, economic and social resources of the American hemisphere, and frankly discusses the best policies for forging the various states into a common opposition to a unified European and Asian zone. Spykman considers opposition to the German-Japanese Axis specifically, but the arguments are applicable to any states seeking to unify their power-region by conquest.

The two parts contain much recent history and good political-geographical descriptions of the regions considered. The maps of hemisphere defence are very useful, but the others are less satisfactory. There is a fair index and a bibliography with some surprising omissions. It is, however, irritating to find so few acknowledgements made in a book which clearly owes a great deal to earlier writers; for example, Ratzel could have been given credit for the concept of three Mediterraneans.

The first two principles described above are a restatement of Spykman's earlier work. The other two were first propounded by Mackinder, who noted the closed nature of the world's political system in 1904, and published his famous prediction in 1919. This prediction ended with the statement, 'Who rules the World-Island commands the World' (Mackinder, 1919, p. 186). Gyorgy (1944) has noted that Spykman was the first American political scientist to accept this postulation and examine its validity.

Reviews of the book reveal a mixed reception. It was hailed by some (Bowman, 1942, and Sprout, 1942), for wakening American political scientists from an idealistic consideration of international relations, to a realistic appraisal which took due account of the facts of geography. On the other hand, the book was criticised (notably by Staley 1942, and Earle 1943) first because it was cast in the mould of *Geopolitik* and ignored the cultural factors in foreign policy; and second, because it provided a dubious plan without showing *how* it could be implemented. It is now proposed to examine these two criticisms in greater detail.

Spykman thought that *Geopolitik* was best defined in the following way (a free translation has been provided):

> Damit Wird Sie Zur Kunstlehre, die praktische Politik bis zur notwendigen Stelle des Absprungs vom festen Boden zu leiten fahig ist.
>
> (Spykman, 1938, p. 30)
>
> Thus it is the technique by which practical politics are based in geographic realities.

Measured by this standard *America's strategy in world politics* belongs to the school of *Geopolitik*. This conclusion is underlined by noting the identity of some of Spykman's ideas with acknowledged members of the German school. According to Haushofer the powerful nations had no cultural or ethical functions to fulfil, and no moral consideration should be allowed to interfere with the state's destiny (Gyorgy, 1944, p. 170). Spykman would appear to agree with this view.

> The statesman who conducts foreign policy can concern himself with values of justice, fairness and tolerance only to the extent that they contribute to or do not interfere with the power objective. They can be used instrumentally as moral justification for the power quest, but they must be discarded the moment their application brings weakness. The search for power is not made for the achievement of moral values; moral values are used to facilitate the attainment of power.
>
> (Spykman, 1942b, p. 18)

Spykman's concept of a world divided into three power-zones in Asia, Europe and America is almost identical with the division suggested by Haushofer-Kjellen (1932, p. 7 ff.) and Dietzel (1940). Haushofer's concept of integrated living spaces, which he called *Grosslebensformen*, bears a close similarity to Spykman's economically integrated region which he called *Grossraumwirtschaft*.

Gyorgy (1944, pp. 253–4) has noted that Spykman's approach to America's problems is similar to a study entitled *Germany prepare for war*, written by Banse (1934). Banse was the most important worker in the field of *Wehrgeopolitik* (geostrategy or defence geopolitics).

Finally two points must be mentioned which modify Spykman's position in the field of *Geopolitik*. The idea of a balance of power, which forms one of the themes of Spykman's book, was strongly attacked by geopoliticians, who regarded it as evidence of the political bankruptcy of the democratic states. The other point concerns a letter which Spykman wrote to the editor of *Life* after that magazine had described him as an 'exponent of cold-blooded power-politics'.

> My interest in a balance of power is not merely inspired by a concern for our power position, but also by my conviction that only in a system of approximately balanced power is collective security possible . . . I am in favour of a balance of power in Europe and Asia, because only in such circumstances can the United States, which is far away, participate effectively in the preservation of international order and undertake positive commitments to preserve the territorial integrity of small states across the oceans.
>
> (*Life*, 11 January 1943)

It does not seem possible to reconcile these views with those set out in the introduction and conclusion of the book.

In examining the second criticism, that Spykman failed to show how his policies could be implemented, the problem seems to have two aspects. First, there is the practical problem of equalising the power-potential of European states by surgery that would witness the disappearance of some small states. This presents difficulties of which Spykman had shown his awareness in his paper on security and international organisation. Second, there is the problem of persuading the United States' citizens to endorse a policy which represents a radical departure from traditions, and which might cut across traditional alliances. In fairness, it must be remarked that Spykman's inability to provide a method of his solution only weakens it, and does not destroy it completely.

The remaining book to be considered—*Geography of the peace*—was published posthumously from material edited by a research assistant. Based as it is upon a lecture, notes and slides made by Spykman before his death, the book is exactly in the style of his earlier contributions. It is a restatement of Spykman's views on America's need to avoid encirclement by a politically organised Eurasian area. There are two introductory chapters, one dealing with theoretical considerations of state security and the other dealing with map projections. The only new suggestion in Spykman's theoretical chapter is that geopolitics has three meanings. To Germans the study provides a framework for history and a doctrine of state expansion, to some, geopolitics is synonymous with political geography, and to Spykman it means the planning of state security in terms of geographic factors. This last suggestion is closest to the proposal of East (1956, p. 23) that the term can be conveniently used to refer to the external geographical relationships of states which affect the whole world. Spykman has clearly changed his views about the definition of *Geopolitik* (*vide supra* p. 37) but has not apparently accepted criticism of the importance he ascribes to geographical factors at the expense of cultural factors. The chapter dealing with the qualities of map projections is valuable in a book which is illustrated by fifty-one maps, the majority being in two colours. The Miller projection is used in most maps. It is a cylindrical projection which combines the accuracy of the Mercator system within latitudes 45 degrees north and south with less distortion of the polar areas.

In the remaining chapters Spykman examines the world strategical views of Mackinder and Haushofer and accepts the former's views, with a slight modification of terminology. The

Heartland is extended westwards to a line linking the Black Sea with Finland. Mackinder's Inner Marginal Crescent becomes the Rimland, and the Outer or Insular Crescent becomes the Offshore Continents. Mackinder's forecast of the pattern of world domination is altered to read, 'Who controls the Rimland rules Eurasia; who rules Eurasia controls the destinies of the world'. This then is the security situation which America must avoid, and the remainder of the book examines some of the ways in which this policy can be effected, some of the problems to be faced, and major events of the Second World War which affected the matter. The final conclusion is that America, Britain and the Soviet Union have most to fear from a unified Rimland, and these three countries should together prevent any other state consolidating its power throughout the region.

Two defects in the book should be noted. Once again there is an absence of references and footnotes, which creates difficulties in placing the book in its subject context. Second, many of the maps are very similar to those used by the German school of *Geopolitik*. Several maps carry arrows of varying size presumably indicating lines of advance or the direction of potential threats, but generally there is no key on the map and insufficient explanation in the text. It is unfortunate that when funds were available for the lavish illustration of the book, they were so badly used.

In summarising this section dealing with the published work of Spykman in the field of international relations two points seem outstanding. Spykman has been rightly praised for putting international politics at the centre of international relations, for awakening American political scientists of the inter-war period to the importance of understanding geographical facts, and for convincingly destroying the arguments of those who would advocate a policy of isolation for America. On the other hand Spykman was correctly criticised for giving too much importance to geography, and ignoring other factors which must contribute to foreign policy.

This view led him very close to the German school of *Geopolitik*, although he did not accept the 'metaphysical nonsense' which was largely window-dressing. It is finally regretted that Spykman seemed unable to profit from the constructive criticism offered by Earle and others.

It has seemed worth while to make a detailed survey of Spykman's published works, since this is the only way in which his final concepts can be understood. Many others, Mouzon, Pounds,

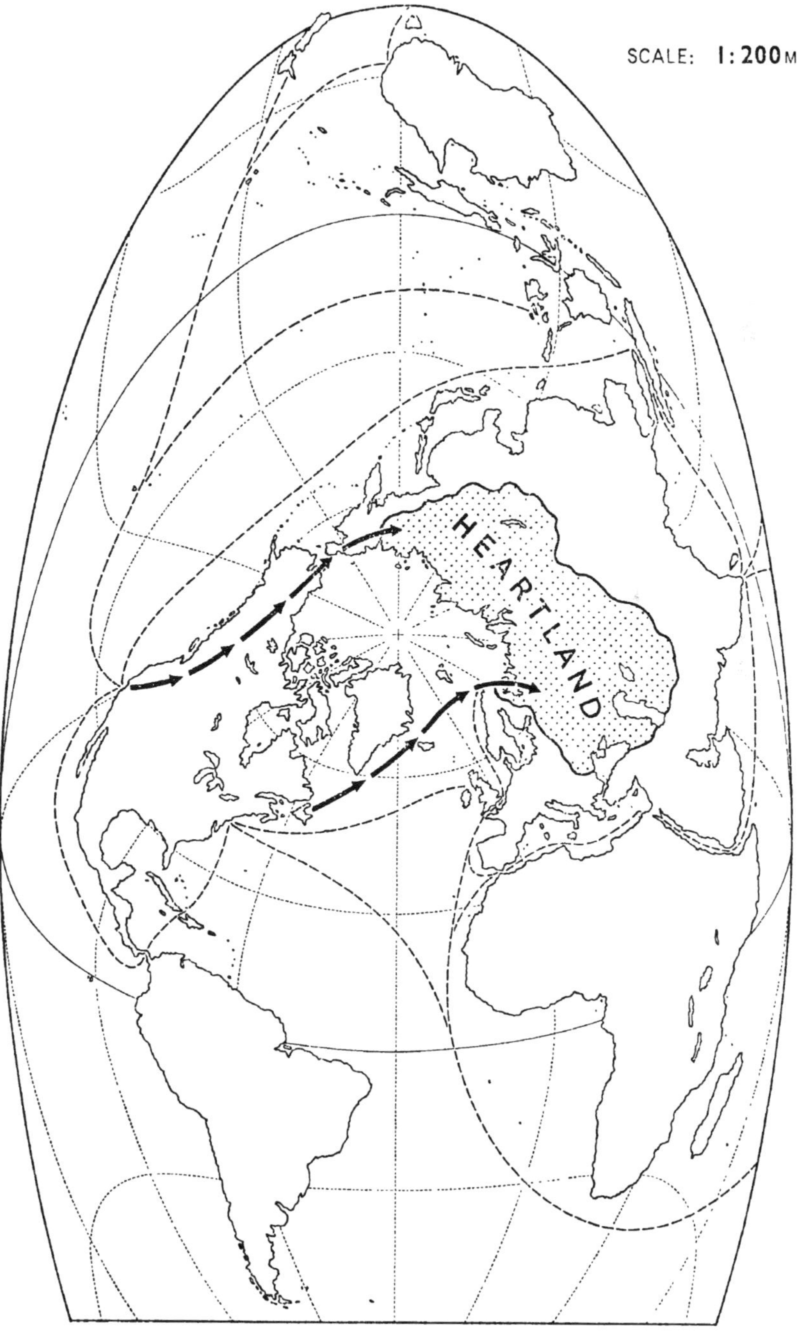

Fig. 1

Geopolitical map of Eurasia after Spykman
Source: *The geography of the peace*
Projection: 'Atlantis' equal area (J. Bartholomew)

Alexander and Cohen included, base their studies of Spykman only on the book edited and published posthumously.

Mouzon has written three contributions to the study of policy. He prepared two reports on petroleum import and trade policies (1952a and 1952b), and published a book dealing with international resources and national policy (1959). It is proposed to concentrate on this book, since it includes most of the material presented in the two earlier reports. The three aims of the book are to provide a textbook to help students interested in the related questions of natural resources and policy, to serve as a guide to policy-making in resource problems, and to recommend a strategy for the security of the United States. After deciding that the main aims of the United States were to preserve freedom and peace and to promote economic progress Mouzon analysed the power basis of the country and made suggestions how that power should be deployed to achieve the stated aims. Eighty-five per cent of this long book is devoted to a consideration of the natural, capital and human resources, which underly the power of the United States. This imbalance is increased by the fact that seventy per cent of the book deals with natural resources, which are considered under the headings of space, agriculture, energy and materials. This unfortunately means that Mouzon is presenting material which has already been substantially considered by geographers, while only introducing the reader to capital resources, which other power analysts have tended to neglect, and which he was well qualified to consider. It is also surprising that Mouzon apparently ignores the theoretical papers which geographers have prepared on national power. For example, there is no mention of the very useful classification of resource availability prepared by Jones (1954). But geographers are not the only group neglected, for there is no reference to the useful studies by the Sprouts (1945 and 1956) on the foundations of national power. There is much interesting information about resources and trade policies, and a wealth of statistics, diagrams and maps, but one must agree with Murdock (1959) that there is too often a gap between the descriptive material about resources and the analyses of policy. The techniques for policy formation are not illuminated by Mouzon, and this is a major disappointment.

In the last section, which aims at discovering how American power can be used to achieve American aims, Mouzon begins with a review of geopolitics for which he accepts Dorpalen's definition.

> Geopolitics is the science of the earth's relationships of political processes. It is based on the broad foundations of political geography, which is *the science of the political organisms in space* and their structure. Moreover, geopolitics sets out to furnish the tools for political action and the directives for political life as a whole.
>
> (Dorpalen, 1942, p. xii, emphasis added)

Later in the book Mouzon quotes Dorpalen's definition of political geography (emphasised above) as a definition of *Geopolitik* (p. 652). It is Mouzon's thesis (pp. 651–3) that past geopoliticians, including Spykman, concentrated on two dimensional space, thereby neglecting resources, which provide the third dimension. This charge can hardly be sustained against Spykman, who devoted two chapters in his main work (1942b, chapters IX and X) to a study of economic patterns and the mobilisation of natural resources. This book is not mentioned by Mouzon. It is also certain that German military strategy took account of the resources and populations of the strategic targets they selected for early conquest. For example, Hitler in *Mein Kampf* wrote of the importance to Germany of the minerals and agricultural wealth of the Ukraine and Urals (quoted in Sprout, 1945, p. 369). The confusion probably springs from Mouzon's conviction that geographic factors include only location, shape and topography, and that agricultural, industrial, mineral and population factors are economic. But political geographers have shown a continuing interest in the political significance of the areal variation of such resources, as the works of Whittlesey (1939) and Van Valkenburg (1939) show. Mouzon's bibliography contains no mention of the multitude of studies by political geographers which might have corrected this impression. Instead he relies entirely on Spykman. This belief in the inadequacy of previous geopolitical studies leads Mouzon to propose a new science—geoecopolitics—which will consist of geopolitics revised to include the neglected economic factors and a greater awareness of the significance of rapid technological change. Mouzon's analysis of geopolitics is made without any reference to the many studies by political geographers, and does little to encourage response to his appeal for contributions from all quarters to 'the inter-disciplinary science in international relations'. The book has deservedly attracted little attention from geographers, although the only geographical review discovered was favourable (Smith, 1960).

Cohen (1964) presents 'a geographical view of contemporary international politics'. The book consists of two main parts. First, Cohen discusses the methodological basis of geopolitical analysis and suggests a division of the world into geostrategic regions. He then describes the power cores of the world and the zones of contact between the major powers.

The main new points made in the methodological introduction concern the importance of the movement factor in power analysis, and his division of the world into geostrategic regions. The movement of men, ideas and goods provides the focal point for studying changes in geopolitical patterns. Such a focus attempts to maintain a contemporary view of geographical realities as they are changed by technological advances, new or altered political philosophies, redistribution of population and new trade patterns. According to Cohen (1964, pp. 16–20) there are three elements in the movement factor: the channel or pathway; the field, which includes origin, route and destination of the channel; and the arena, which is the space—land, sea or air—within which the channel and the field lie.

Because Cohen's concepts of geostrategic and geopolitical regions are not easy to understand, it is important to describe them in his words.

> The geostrategic region must be large enough to possess certain globe-influencing characteristics and functions . . . [it] . . . is the expression of the inter-relationships of a large part of the world in terms of location, movement, trade orientation, and culture or ideological bonds. While it is a single-feature region, in the sense that its purpose is to embrace areas over which power can be applied, it is a multi-feature region in its composition. Control of strategic passageways on land and sea is frequently crucial to the unity of geostrategic regions.
>
> The geopolitical region is a subdivision of the above. It expresses the unity of geographic features. Because it is derived directly from geographic regions, this unit can provide a framework for common political and economic actions. Contiguity of location, and complementarity of resources are particularly distinguishing marks of the geopolitical region . . . the geostrategic region has a strategic role to play and the geopolitical region has a tactical one.
>
> (Cohen, 1964, p. 62)

This passage contains a number of phrases which present problems of interpretation. 'Globe-influencing characteristics and functions' is taken to mean that a defined portion of the world possesses qualities and contains operations which influence both that area and the rest of the world. But surely this statement does not give proper acknowledgement to the role of governments. It is the actions of governments which frequently imparts a particular significance to regional qualities; operations within any region are conducted in large measure under the aegis of the governments which control the area; and it is the reaction of governments, in other parts of the world, to these qualities and operations which are important. It must also be noticed that events in very small areas, such as the Suez Canal, can produce reactions throughout much of the rest of the world.

The next part of the concept is simpler to understand, but the problem of interpretation arises from the need to weight the interrelationships which are specified. For example, is trade orientation more important than ideology, or less important than location? Geographers giving different values to these relationships would construct different maps of major world divisions.

The idea that the geostrategic region defines the area over which power can be applied depends for its interpretation on the meaning given to the term 'power'. Does it mean actual control or significant influence? Can it only be wielded by a formal political organisation such as a single state or an association of states? The reference to the unity of the geostrategic region being influenced by control of vital routes connotes a certain measure of political control by a single state or alliance of states, which presumably must be part of the geostrategic region. It seems likely that control could also be interpreted as the absence of obstructions or restrictions. Neither the American nor the British Navy controls the sea lanes linking the Trade-Dependent Maritime World, but neither does the navy of any other friendly or hostile state.

The statement that the geopolitical region expresses the unity of geographic features does not serve to distinguish it from the geostrategic region, for location, movement, trade orientation and culture are geographic features, in the sense that their quality varies over the earth's surface. The belief that contiguity of location and complementarity of resources are two of the distinguishing marks of the geopolitical regions raises two points apart from the question of the weight to be given to each. First, there is the implication that states in the same general area will have the same general strategic and economic view of the world.

There are plenty of examples in Asia, Europe and Africa to demonstrate that this is not the case. Second, it is not clear whether the complementarity of resources refers to developed or potential resources, in actual or theoretical political circumstances. Given the present level of development and the political dichotomy of Africa, the continent does not form a geopolitical region in the sense of complementarity of resources.

It is difficult to understand how either of the regions can play any sort of strategic or tactical role *by themselves*. It might be accepted that different governments will decide that they have an overall (strategic) interest in certain major regions of the world, and specific (tactical) interests in subdivisions; but some of Cohen's geopolitical regions seem too large to describe interest in them as tactical, even by comparison.

These problems of interpretation suggest that the scheme needs much more clarification if it is to prove useful to geographers interested in international relations. This point is underlined by the fact that even if it is accepted that it is a useful exercise to divide the world in this way, there are details of Cohen's map which must be questioned. Cohen defines two geostrategic regions, the Trade-Dependent Maritime World and the Eurasian Continental World, each of which is subdivided into a number of geopolitical regions. There is a geopolitical region in the Indian sub-continent, which is distinct from both geostrategic regions and the flanking shatter belts. The two shatter belts, South-east Asia and the Middle East, are the areas occupied by a number of conflicting states, in which the major powers have conflicting interests. It is difficult to understand why the Sudan and Libya are both divided between the Trade-Dependent Maritime World and the Middle East shatter belt. Surely the boundaries of the geopolitical and geostrategic regions ought to coincide with international boundaries? The scale and distortion of Cohen's map make it impossible to know whether he has followed international boundaries on the western and southern edges of the Eurasian Continental Power geostrategic region.

At various points in his book Cohen advances reasons for making this kind of study.

> The purpose of this book is to present a geographical view of contemporary international politics.
>
> (p. iii)

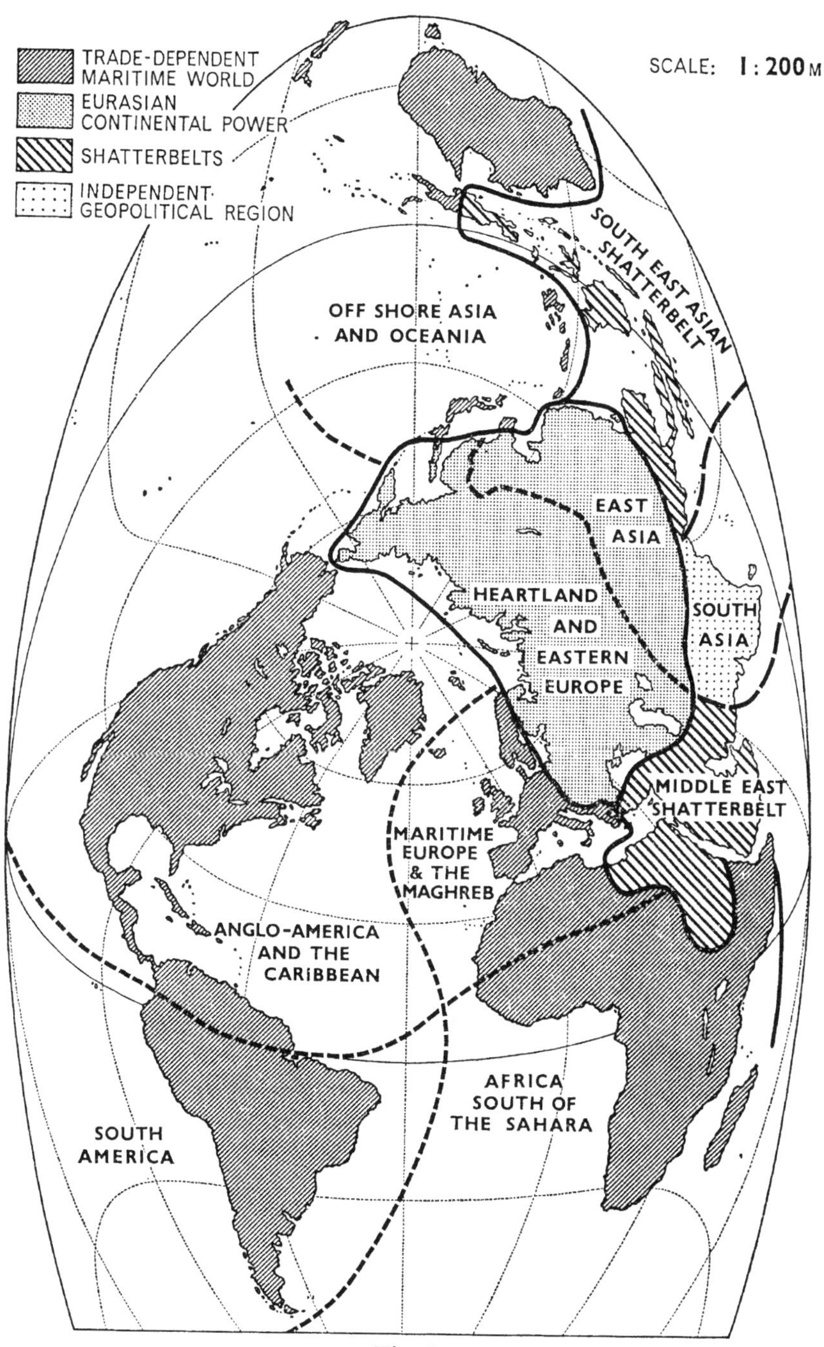

Fig. 2

The World's geostrategic regions and their subdivisions after Cohen
Source: *Geography and Politics in a divided World*
Projection: 'Atlantis' equal area (J. Bartholomew)

What purpose does geopolitical analysis serve? Harold and Margaret Sprout feel that such hypothesising 'may serve purposes of policy-making and propaganda, and that whatever the avowed interests of the authors . . . [their geopolitical writings] have tended to serve both kinds of purposes . . . Granting the truth of the observation made by the Sprouts, we would simply add that the geopolitical analyst is validly fulfilling this dual function so long as he does not deliberately distort the geographical setting *as he sees it* and as long as he does not lay claim to being a practitioner of an empirically based science.
(p. 25)

Geopolitical analysis has two major aspects: (1) description of geographical settings as they relate to political power, and (2) laying out of spatial frameworks that embrace interacting political power units.
(p. 25)

This is indeed the difficulty with political regions. They are real and tangible for the moment, but if they lack firm groundings in broader political, social, economic and physical 'realities', then they are fleeting. It is to geography that we turn for a true appreciation of political realities. The geographical setting, both that which is fixed and that which is dynamic, provides us with a basis for understanding today's political map and for anticipating change. Therefore the geopolitical map is more closely attuned to reality than the political map.
(p. 62)

The first and fourth points are similar, and usefully stress the need for political scientists to show a greater awareness of geographic factors. This reference to political realities echoes a point made by Jones (1959) when he drew attention to the early efforts of Langhans (1926) to show the degree of self-government in various parts of the world. Langhans experienced difficulties which derived partly from his subjective views of history and partly from the differences between legal and practical situations in some territories. This point also relates to the earlier discussion that parts of the state outside the control of the government do not form *de facto* parts of the state. The present political boundaries of former Indo-China do not reflect the realities of political control. The second point can be accepted, although it would be helpful to provide some clearer indication of how significant this general world survey is to the formation of policy. It seems likely that it might be important in the initial stages, but would soon give way to more specific studies. It might also be worthwhile to insist that political geographers should not deliberately serve propaganda purposes, although it must be recognised that even the most objective study may support one view rather than another, and may be capable of misquotation for propaganda purposes.

There are two other striking features of Cohen's style: his personal involvement, and his readiness to advocate policy.

> . . . our sea lanes (p. 57) . . . our global security (p. 57) . . . our ventures in the Middle East (p. 59) . . . we invoked the Eisenhower Doctrine (p. 60) . . . our ability to work out accommodations with these non-Arab states (p. 60) . . . we have buttressed France's position (p. 132) . . . our relations with Latin America (p. 133) . . . we fear counter-encirclement from the Soviet Arctic (p. 133) . . . our global foreign policy.
>
> (p. 139)

This style of writing, together with frequent references to the 'Free World', is not consistent with objective political geography, and may have been the reason why Burghardt (1956), whose help is acknowledged by Cohen in the Preface, suggests that Cohen 'wishes to impress his view of the world geopolitical situation on the leaders of the United States'.

There are many instances where Cohen recommends a particular policy.

> What we cannot afford to do, however, is to take the American Mediterranean for granted.
>
> (p. 138)

> As long as its most important allies are so heavily dependent upon overseas trade the United States has to help them maintain their sea contacts.
>
> (p. 66)

> Today we are in the danger of going to the other extreme and of disengaging ourselves from too many areas.
>
> (p. 285)

> In such an event [the failure of Europe to reach accommodation with Black Africa] all of Africa south of the Sahara could become a shatter belt, within which the maintenance of footholds in the mineralised Highland South and in West Africa, would become the minimum strategic requirements of the Maritime World.
>
> (p. 277)

> Because partition [of Algeria] is no longer a realistic alternative, France and the West have no alternative but to make sure the Evian Agreements are upheld.
>
> (p. 184)

Burma, also, warrants our wholehearted protection against Communist encroachments.

(p. 263)

Cohen might be right in every one of these judgements, but it must be recognised that they cannot be based on geography alone, there are political, economic and moral issues which should be taken into account, and which the political geographer has no special qualification to consider. If political geographers are going to recommend policy they should invariably add the rider that non-geographical factors may make such a proposal unsuitable.

This review of three global strategies suggests three major conclusions. First, despite the criticisms which have been advanced against these various studies, there is a need for global views and global strategies. This point has been stressed separately by distinguished political geographers.

Geopolitics, despite the ill-famed *Geopolitik* of the inter-war years, remains a valid field for the political geographer in search of a difficult and useful exercise; it remains one of the few fields where the geographer still dares to look the whole world squarely in the face.

(East, 1965, p. 417)

. . . men must adopt systems of thought that sift and order the vast amount of communication that assaults the senses every day . . . such thought-filters are systems of evaluation as well as mere ordering, and therefore may cause distortion . . . that distortion is inevitable when some one aspect of the world is permitted to dominate our view of it. The world is much too complex to be evaluated in terms of a single aspect. There is a basic paradox here: the world is so complex that we must systematise our knowledge of it, but the very complexity of the world makes our systems necessarily imperfect. The important thing of course is to remember that our global views are imperfect, to use them as aids to thought but not as dogmas.

(Jones, 1959, p. 67)

The way in which an event in one part of the world can cause chain reactions in economic development, population movement, capital flow and political influence in other widely separated areas is further proof of this need.

Second, these studies occupy the meeting point of a number of disciplines which includes political science, economics and geography. It follows that co-operation amongst members of different disciplines is likely to produce the most complete and accurate picture of global strategical realities. This point also means that workers in one or another field should avoid the temptation to strongly advocate policy on the basis of their own research alone.

Third, while members of each discipline have a duty to co-operate with the others they also have the responsibility to develop, to the maximum extent, those parts of the field in which they have the greatest competence. In the case of political geography the greatest benefit would seem to flow from theoretical studies such as those of Jones (1955 and 1959) and studies which present the significance to policy-making of geographical factors. The view expressed earlier that there are as many global views as there are countries is clearly illustrated by the differences in quality between world surveys written by one author and many authors. The studies by Cohen and Spykman particularly reflect the nationality and viewpoint of the author, in the contrast between their detailed treatment of America and the oversimplification of certain other areas. Alexander (1957) made a more satisfactory attempt to present a world survey, but it is impossible for one man to write with uniform authority and insight on all parts of the world. The ideal consists of contributions from many authors; and the best examples are those studies by East and Moodie (1956) and East and Spate (1961). The disadvantage of such surveys is that it may be difficult to have the book regularly revised. It would also be helpful if there were more regional studies involving a number of countries, for these seem to be a valid subdivision of the global view. Some of the best regional studies have been made by Fisher (1950, 1963 and 1965). It will also be useful for geographers to produce a number of studies of the same area at different times, since this will give a useful insight into the way in which geographical factors change in significance, as technical developments occur in transport, agriculture and manufacturing.

REFERENCES

ALEXANDER, L. M., 1957, *World political patterns*, Chicago.

ANCEL, J., 1938, *Les frontières*, Paris.

BANSE, E., 1934, *Germany prepare for war: a Nazi theory of 'National Defense'*, New York.

BOWMAN, I., 1921, *The new world: problems in political geography*, New York.

BOWMAN, I., 1942, 'Political geography of power', *Geogr. Rev.*, **32**, pp. 349–52.

BRUHNES, J., and Vallaux, C., 1921, *Géographie humaine de la France*, Paris.

BURGHARDT, A., 1965, 'The dimensions of political geography: some recent texts', *Canadian Geographer*, **9**, pp. 229–33.

COHEN, S. B., 1964, *Geography and politics in a divided world*, New York.

COHEN, S. B., 1966, 'Toward a geography of policy', *Econ. Geogr.*, **42**, Guest editorial for January.

DIETZEL, K. H., 1940, 'Imperialismus und Kolonialpolitik', *Zeitschrift für Geopolitik*, **17**, pp. 313–22 and 372–6.

DORPALEN, A., 1942, *The world of General Haushofer*, New York.

EARLE, E. M., 1943, 'Power politics and American world politics', *Amer. Pol. Sc. Quart.*, **58**, pp. 94–105.

EAST, W. G., 1937, 'The nature of political geography', *Politica*, **2**, pp. 259–86.

EAST, W. G., 1950, 'How strong is the Heartland?', *Foreign Affairs*, **29**, pp. 78–93.

EAST, W. G., 1965, review of *Geography and politics in a divided world*, *Geogr. J.*, **131**, pp. 417–18.

EAST, W. G., and Moodie, A. E., Eds., 1956, *The changing world*, London.

EAST, W. G., and Spate, O. K. H., 1961, *The changing map of Asia*, 4th ed., London.

FEBVRE, L., 1925, *A geographical introduction to history*, New York.

FISHER, C. A., 1950, 'The expansion of Japan: a study in Oriental Geopolitics', *Geogr. J.*, **115**, pp. 1–19 and 179–93.

FISHER, C. A., 1963, 'The Malaysian Federation, Indonesia and the Philippines: a study in political geography', *Geogr. J.*, **129**, pp. 311–28.

FISHER, C. A., 1965, 'The Vietnamese problem in its geographical context', *Geogr. J.*, **131**, pp. 502–15.

GYORGY, A., 1944, *Geopolitics: the new German science*, University of California publications in International Relations, **3**, pp. 141–304.

HARTSHORNE, R., 1935, 'Recent developments in political geography', *Amer. Pol. Sc. Rev.*, **29**, pp. 758–804 and 943–66.

HARTSHORNE, R., 1936, 'Suggestions on the terminology of political boundaries', *Annals*, Association of American Geographers, **26**, pp. 56–7.

HARTSHORNE, R., 1954, 'Political geography' in *American Geography, inventory and prospect*, James, P. E., and Jones, C. F., Syracuse.

HAUSHOFER, K., and Kjellen, R., 1932, *Jenseits der Grossmachte*, Berlin.

HENNIG, R., 1931, *Geopolitik*, Leipzig.

HINKS, A. R., 1921, 'Notes on the techniques of boundary delimitation', *Geogr. J.*, **58**, pp. 417–43.

JONES, S. B., 1954, 'The power inventory and national strategy', *World Politics*, **6**, pp. 421–52.

JONES, S. B., 1955, 'Views of the political world', *Geogr. Rev.*, **45**, pp, 309–26.

JONES, S. B., 1955, 'Global strategic views', *Geogr. Rev.*, **45**, pp. 492–508.

JONES, S. B., 1959, 'Global strategic views', in *Military aspects of world political geography*, United States Air Force, Alabama.

KIRK, W., 1962, 'The inner Asian frontier of India', *Transactions*, Institute of British Geographers, **31**, pp. 131–68.

LANGHANS, M., 1926, 'Karte des Selbstbestimmungrechtes der Volker', *Petermanns Mitteilungen*, **72**, pp. 1–9.

LAPRADELLE, P. de, 1928, *La frontière*, Paris.

MACKINDER, H. J., 1904, 'The geographical pivot of history', *Geogr. J.*, **23**, pp. 421–44.

MACKINDER, H. J., 1919, *Democratic ideals and reality: a study in the politics of reconstruction*, New York.

MAULL, O., 1925, *Politische Geographie*, Berlin.

MORGENTHAU, H. J., and Thompson, K. W., 1956, *Principles and Problems of international politics*, New York.

MOUZON, O. T., 1952a, *The United States Petroleum trade policy*, Washington.

MOUZON, O. T., 1952b, *U.S. petroleum import policy*, Washington.

MOUZON, O. T., 1959, *International resources and foreign policy*, New York.

MURDOCK, J. C., 1959, review of *International resources and national policy*, *Amer. Econ. Rev.*, **49**, pp. 1120–1.

RATZEL, F., 1895, *Politische Geographie*, Munich.

RYDER, C. H. D., 1925, 'The demarcation of the Turco-Persian boundary in 1913–14', *Geogr. J.*, **66**, pp. 227–42.

SMITH, G. H., 1960, review of *International resources and national policy*, *Economic Geography*, **36**, pp. 375–6.

SOLCH, J., 1924, *Die Auffassung der 'natürlichen Grenze' in der wissenschaftlichen Geographie*, Innsbruck.

SPROUT, H. H., 1942, review of *America's strategy in world politics*, *Amer. Pol. Sc. Rev.*, **36**, pp. 956–8.

SPROUT, H. H., and M., 1945, *Foundations of national power*, New York.

SPROUT, H. H., and M., 1956, *Man-milieu relationship hypotheses in the context of international politics*, Princeton.

SPYKMAN, N. J., 1933, 'Methods of approach to the study of international relations', *Proceedings of the fifth Conference of Teachers of International Law and Related Subjects*, pp. 60–9, Washington.

SPYKMAN, N. J., 1938, 'Geography and foreign policy', *Amer. Pol. Sc. Rev.*, **32**, pp. 28–50 and 213–36.

SPYKMAN, N. J., and Rollins, A. A., 1939, 'Geographic objectives in foreign policy', *Amer. Pol. Sc. Rev.*, **33**, pp. 391–410 and 591–614.

SPYKMAN, N. J., 1942a, 'Frontiers, security and international organisation', *Geogr. Rev.*, **32**, pp. 436–47.

SPYKMAN, N. J., 1942b, *America's strategy in world politics*, New York.

SPYKMAN, N. J., Ed. Nicholl, H. R., 1944, *The geography of the peace*, New York.

STALEY, E., 1942, review of *America's strategy in world politics*, *Amer. Econ. Rev.*, **32**, pp. 457–61.

VAN VALKENBURG, S., 1939, *Elements of political geography*, New Jersey.

WEIGERT, H. W., 1949, 'Mackinder's Heartland', in *New Compass of the world*, Weigert, H. W., Stefansson, V., and Harrison, R. E. (Eds.), New York.

WHITTLESEY, D., 1939, *The earth and the state*, New York.

3

POLICIES FOR THE DEFENCE OF THE STATE

Policies considered in this chapter are designed to insure the state against any actual or potential threat to its territorial extent or established areas of interest. Threats to the territorial integrity of a state will normally take one of two forms. First, there is the case of internal disruption, when the population of part of the state wishes to secede, either to join another state, or to create a new independent state. Such action may, or may not, be encouraged by alien states. The claims of the Eastern Region of Nigeria to independence as the Republic of Biafra in May 1967 provided an example of secession to create a new state; the Shifta campaign of Somali in north-east Kenya, designed to link their territory with the Somali Republic, is an example of the second instance. The second form results from the external threat of a neighbouring state which seeks to annex all or part of the territory. The post-war world abounds with examples of this situation: the efforts of North Vietnam to unify the whole of Vietnam, the efforts of Indonesia to win West Irian; and the efforts of India to annex Goa and Daman.

Although the final result in each case will be the same to the state concerned—that is the loss of territory—it is worth separating these two cases because of the different counter policies which may be employed. In the case of an internal threat, the state may take action which seems appropriate, providing it still controls the area in question. It may attempt to coerce the population by military campaigns, such as those launched by various Irak Governments against Kurdish independence movements, or by economic pressures, which was the first method employed against Biafra by the Nigerian federal authorities. Alternatively the government may attempt to destroy the dissidence by removing its cause. This may be done by improving economic conditions, or by removing constitutional disabilities, or by granting a greater measure of autonomy, as India has done in the case of certain hill states such as Nagaland. These remarks apply not only to the contiguous areas of the state, but also to any overseas possessions which it may control. Thus the Rhodesian declaration of independence in November 1965 was a form of secession which was countered by economic policies of coercion. In the case of external threats, a government is limited in the actions it can take, because it lacks control over the area from which the threat originates. This will not prevent the state from initiating unilateral economic or military measures, as the United States has taken action against North Vietnam, but equally, the state may employ more oblique diplomatic measures to neutralise the danger.

With these policies designed to meet the internal and external threats, which in some cases may be combined, it seems worthwhile to include also policies designed to regain territory lost through the actions of a dissident minority or the activities of another state. Such policies represent an attempt to redress the failure of earlier policies, which failed to counter the threat. The efforts of the Governments of Syria, Jordan and the United Arab Republic to regain territory captured by Israel in June 1967 furnish a recent example. Other cases are provided by Germany's struggles, during the period 1930–45, to regain certain areas which had been detached during the peace settlements at the end of the First World War, and the recent efforts of Spain to win back Gibraltar, which was ceded, under some pressure, to Britain, by the Treaty of Utrecht in 1713. Such policies may be judged to be aggressive in terms of the existing territorial arrangement, but it is essential to realise that the terms 'defensive' and 'aggressive' are only relative, and that policies designed to defend the state may result in the acquisition of additional territory. This situation is exemplified by Russia's gains in Europe during and after the Second World War, by Israel's gains at the expense of the Arab states in 1967, and by Germany's gains at the expense of France in 1871. Policies of acquisition have been common at various times in history, when a state felt that its legitimate interests, often commercial, were being threatened or infringed. The spur of competition by another state's nationals was often decisive in the decision of an imperial power to formalise *de facto* control, which had carried all the advantages and none of the responsibilities. Britain's annexation of the Natal coast and Bechuanaland, in order to isolate the Boer Republics from direct contact with alien states, provides a good example of this situation.

It must also be recognised that the state may be subjected to external threats which are not designed to win territory, but which seek concessions in such spheres as trade, communication, the treatment of minorities and immigration. This represents a curtailment of the government's power to act, rather than its territory, but such infringements will also be met by policies of defence.

An examination of policies for the defence of the state suggests that the most meaningful division separates unilateral and multilateral policies. The significance lies in the need to construct a single or multiple view of the significant factors. It is often the case that the unilateral policies will be applied within the boundaries of the state, but this is not an inviolable rule; it is also possible for a state to take unilateral action outside its boundaries, or in such a way that it infringes the rights of another state. War often starts by unilateral action, as Israel has demonstrated on two major occasions. The corollary is that multilateral policies need not necessarily be applied to areas outside the state. A government may reach bilateral or multilateral agreements which will apply to the territory of the state itself. Such policies would include the acceptance of finance to build strategic roads or airstrips, the granting of base facilities to another state, and the acceptance of troops from another state. In many cases, however, multilateral agreements will be concerned with areas outside the state, or with relationships between states. In some cases the respective governments will have a legal right to conclude agreements dealing with defence, in areas such as trade in strategic materials or immigration restrictions; in other cases, as when two states agree to partition all or part of a third, the action is based on *de facto* power rather than *de jure* right.

These thoughts suggest a threefold division of policies for the defence of the state. First, unilateral policies which are applied to areas where the state has authority and legal prescription, which may be called policies of organisation. Second, multilateral policies which may be legally concluded, and which may be styled policies of diplomacy; and third, multilateral or unilateral acts of a violent nature, which may be described as policies of conflict. Such a classification would be based on grounds of convenience, to clarify analysis, for it must be stressed that there is a wholeness about policy-making. Policies of organisation may be designed to strengthen the state's position before launching policies of diplomacy or conflict. Policies of conflict will sometimes begin when policies of diplomacy fail, although there is no obstacle to both proceeding at the same time. For example, during the Vietnam war, America has maintained a diplomatic offensive, and during the British blockade of Rhodesia, officials of both governments, even the Prime Ministers, met to search for a solution.

It is clear that in all these types of policy of defence geography may be significant at three distinct stages. First, geographical factors may be relevant to the initiation of a particular policy, or its re-emphasis. For example, the decision of the Transvaal Government to concentrate on securing the rail link with Delgoa Bay, before admitting any lines from the Cape Colony, followed the discovery of gold in the Rand. A few months previously, the adverse economic situation had decided President Kruger to seek an economic agreement with the Cape Colony; these overtures were rejected because it seemed that the weakness of the Transvaal would ultimately lead to greater concessions. Second, geographical factors may influence the conduct of a particular policy; this would include the overall strategy as well as the detailed tactics. Such geographical factors will be most obvious in the case of policies of conflict, but they may also influence the place and pace of diplomatic activity, or methods of organisation. Lastly, geographical factors may influence the decision to end certain policies or to convert them to another form. For example, major losses of territory may induce a government to abandon policies of conflict for policies of diplomacy to avoid further losses; this situation faced the Mexican Government in 1848 and the Arab Governments in 1956 and 1967.

All the policies for the defence of the state are likely to have one or both of two primary aims—to make the state and its allies stronger, and to make the opponents of the state weaker. Policy decisions to strengthen the state will usually aim at either an improvement in the state's intrinsic power, or the acquisition of allies. Policies to weaken a hostile state will usually centre on attempts to increase their problems, truncate their military and economic bases, and deny them the assistance of allies.

The elimination or exploitation of dissidence

One method of increasing the intrinsic strength of a state is to reduce areas of dissidence where they occur; conversely, the exploitation of dissident groups in a hostile state will tend to weaken that state by increasing its problems. Dissidence will normally occur amongst a section of the population which regards itself as being distinctive in some way, such as language, religion, tribal origin or commercial interest. In many cases these sections will be associated with a particular area. Groups such as the Somali of Kenya, the Kurds of Iraq, Iran and Turkey, and the Nagas and Mizos of Assam spring to mind. Any government faced with a dissident section can attempt to eliminate the problem by persuasion, or by force including surveillance, or by expulsion. Attention is being focussed here on groups seeking to secede or help an alien enemy state, either directly or by non-co-operation, rather than groups seeking to overthrow an existing government by revolution, an act which will not usually involve territorial change. Secessionist groups are generally more common than sections of the population who will either aid an enemy or refuse to promote the welfare of the state. These latter groups may include political parties, aliens, trade unions, or fringe religious sects. Such groups rarely occupy a contiguous area, but draw their membership from various parts of the state and from various social groupings.

The efforts of various Iraq Governments to solve the Kurdish question illustrate policies of coercion and persuasion. The Kurds live in the borders of Turkey, Iraq and Iran; they have a feeling of affinity for each other and generally speak one of two main Kurdish dialects—Zaza in the north and Kermanji in the south (Arfa, 1966, chapter 1). The division into the present states occurred after the First World War. Edmonds (1967) has provided a very useful and concise account of the development of the Iraq Kurdish question. According to the Treaty of Sèvres in 1920, Turkey and the Allies agreed to the creation of autonomous Armenian and Kurdish areas. This treaty was never ratified because of the actions of Kemal Attaturk. It was replaced in 1923 by the Treaty of Lausanne, which made no mention of Armenia or Kurdistan; but this treaty stipulated that the future of the Mosul *vilayet* (a subdivision of the Ottoman Empire) should be decided by direct negotiations between Turkey and Britain, as mandatory power for Iraq.

These countries were unable to agree and the matter was

referred to the League of Nations, which ruled in Iraq's favour in 1926,

> . . . on condition that regard should be had to the desires expressed by the Kurds that officials of Kurdish race should be appointed for the administration of their country, the dispensation of justice, and teaching in schools, and that Kurdish should be the official language of all these services.
>
> (Quoted in Edmonds, 1967, p. 11)

In 1931 the Iraq Government enacted the Local Languages Law, which nominated districts in which Kurdish might be used for official purposes.

The political geographical significance of the area occupied by the Kurds derived partly from its position on the borders of Iran and Turkey, with whom Iraq wanted good relations; partly from the fact that one of the main routes to Iran lay through the Kurdish areas of Diyala; and partly from the fact that petroleum had been discovered, in 1927, at Kirkuk and Jambur, within Kurdish-dominated areas.

Up to the end of the Second World War there had been various uprisings by sections of the Kurds. In 1932, Shaikh Ahmad forcibly resisted a British plan to settle Nestorian Christians, expelled from Turkey, in the vicinity of Barzan. In 1929 and 1930 there were some risings in favour of Kurdish autonomy around Sulaymaniyah; and in 1945 there was a major rising throughout northern Iraq, sponsored by the Kurdish Nationalist Party, with the support of religious and tribal leaders. The policy adopted by the administrations to meet all these challenges was military conquest. The pattern of military campaigns had a similarity which owed much to the nature of the country. The government forces would seek to hold the lowland communications and the principal settlements while the Kurdish forces withdrew to the hilly, less accessible areas, from which they launched raids on the government's extended lines of communication. Campaigns were normally discontinued during the winter because of the deep snow which covered many areas. Increasingly the governments sought to make use of planes and artillery to counter the greater mobility of the Kurds. These weapons were used to destroy rebels' crops and villages. During these periods of strife there were frequent short- and long-term population movements; the 1945 campaign was inconclusive because the insurgents retreated into the Kurdish areas of Soviet-controlled Iran. It was also the policy of the governments to employ loyal Kurds to fight against the rebels. This had the effects of employing a force which could use the terrain as well as the rebels, and of reducing the level of Kurdish hegemony. These loyal tribes included the Lowlan, the Herki Berati and the Baradost (Arfa, 1966, pp. 140 and 145). There have been further disturbances in 1961, 1963 and 1965, interspersed with negotiations which have so far (June 1968) proved fruitless; the main areas of dissidence have included Barzan, Agrah, Margasur and Amadigah. It is interesting that the three rebellions were against three different governments, led by Abd-el-Karim Qasem, the Baath Party, and the Aref brothers respectively. In each case the new government started with friendly relations, which then deteriorated to a point where the decision was taken to attempt to control the Kurds by a military campaign. The demands of the Kurds have been fairly constant, centring on a significant measure of autonomy, the employment of Kurds as government officials throughout Kurdistan, the use of Kurdish as a joint official language, representation in the central government, and a share of oil revenues. Governments have promised to meet most of these demands, except for the sharing of oil revenues, but they have not implemented them at a rate considered satisfactory by some of the Kurdish leaders.

For example, the Baathist Government proposed a scheme of decentralisation, which arranged for the division of Iraq into six regions (Muhafizas), called Mosul, Kirkuk, Sulaymani, Baghdad, Hilla and Basra. The boundaries of these regions did not coincide with the boundaries of the former Turkish *liwa* (districts), but followed a combination of various district and sub-district boundaries. In Sulaymani, where the bulk of the population was Kurdish, their language would rank with Arabic as the official medium. The alignment of the boundary, however, excluded certain sub-districts in Kirkuk, Baghdad and Mosul, where the Kurdish population in 1956 represented at least 70 per cent of the district total. These areas included the principal route to Iran via Khanaquin and the major oil fields.

These population proportions will no longer apply in the sub-districts, because of a deliberate policy of replacement of Kurds by Arabs, which the last two governments have pursued. Edmonds (1967, pp. 17–18) has noted this, and the Kurds have alleged that thirty-seven Kurdish villages in the Erbil oil-producing region were evacuated and replaced by Arabs. The present government (June 1968) arranged the last cease-fire in June 1966, and since then there have been fitful discussions on the government's concessions. The most significant of these for the political geographer include recommendations on a degree of autonomy within Kurdish districts (rather than *one* district) in respect of health, education and municipal affairs; Kurdish representation in parliament and the cabinet according to the proportion of Kurds in the total population of Iraq; the recognition of Kurdish as an official language; and the formation of a Ministry for the reconstruction of Kurdistan.

Some interesting cases revealing policy attempts to deal with potential secessionist movements have occurred during the process of decolonisation. The constitutions of Nigeria, Cyprus, Uganda and Kenya were all designed to promote national unity, which was clearly in jeopardy judged by the statements of certain leaders of minority groups represented in the constitutional discussions. In the cases of Nigeria, Kenya and Uganda this involved a constitutional and territorial arrangement, which was made possible because the various tribal factions were located in discrete areas. In Cyprus, due to the complex intermingling of the two main ethnic groups, Turkish and Greek Cypriots, no territorial arrangement was possible except in the major cities, where there were distinct ethnic quarters; instead guarantees were written into the constitution. The 1959 Treaties of Zürich and London laid down certain rights for the Turkish minority, which forms about 18 per cent of the population. These measures included a Turkish vice-president with the power of veto over certain spheres of action, including foreign affairs, defence, and security; three Turkish members in the ten-man executive; and fifteen Turkish members of the fifty-man House of Representatives. These fifteen members had the right of a separate vote on fiscal and municipal matters, and a two-thirds vote was necessary for the passage of any bill; thus six Turkish members could block any legislation. Thirty per cent of positions in the public service and 40 per cent of positions in the army and the police were also reserved for Turkish Cypriots.

These rights, in effect, gave the minority the right to stultify legislation, and for three years Cyprus had no taxation or municipal laws! Proposals by Archbishop Makarios to modify these conditions in November 1963 resulted in a crisis and the establishment of a United Nations peace-keeping force. There has also been some redistribution of the population, as some Turks have withdrawn to areas of denser Turkish settlement. Some of the main cities, including Nicosia, are divided into Turkish and Greek Cypriot sectors by so-called 'green lines', which are dividing zones manned by United Nations army and police. Another effect of the crisis was seen in the reduction in tourist earnings.

	1963	*1964*	*1965*
Earnings from tourism (Millions £ sterling)	4·5	1·0	2·3
Number of tourists	74,619	16,084	33,246

(Source: *Europa Year Book*, 1966, Vol. I)

The significance of this decline is made more apparent when it is realised that in 1965 only the export of oranges and cupreous concentrates earned more, and not much more, than the tourist industry.

The Indian Republic has been beset with a number of secessionist movements since its federation, and on 16 June 1966 the Central Government passed an ordinance specifying penalties for individuals or groups who advocated the secession of any part of India. In the previous six months there had been demands for independence from certain groups of Sikhs in the Punjab, and some parties in Kashmir, Nagaland and the Mizo Hill Area. The Indian Government has usually met such demands with military firmness and proposals for limited autonomy. For example, the Sikhs have been awarded a separate Punjabi-speaking state through the partition of the Punjab; in similar fashion the area of Nagaland has been set up as a state within the federation. The Hill Areas of Assam—Mizo, Khosi Jaintia, Garo and Mikir and North Cuchar—have provided many administrative difficulties. The areas are located in a *cul-de-sac* of Indian territory, bounded by Burma and East Pakistan, communications with Assam and the rest of India are poor, and the country is hilly and thickly forested. The population has certain distinctive cultural traits in language, and often religion. For example, 80 per cent of the 200,000 Mizos are Christians, and the level of literacy (45 per cent) is the second highest in India. Some of these people were in armed rebellion in March 1966, which the government at first countered by force, and then by the proposal to create a federal state of Assam, in which the Hill Areas will form either one or four units. This would be a federation within a federation. The statements made at the time showed the government's concern with the location of this region on the borders of Burma and East Pakistan, close to the area of the North-east Frontier Agency, which is contested by the Chinese Government.

> The Government of India appreciates the political aspirations of the people of the Hill Areas of Assam, and has decided to recognise the State of Assam, bearing in mind the geography and the important needs of security and coordinated development of this region as a whole.
> (*Keesing's Contemporary Archives*, 1967, p. 21, 852)

This concern must have deepened as a consequence of disturbances in the narrow neck of land between Nepal and East Pakistan, in July 1967.

The examples considered so far were countered by policies of persuasion and coercion; it is now necessary to look at policies designed to give the government control over the situation by increased surveillance, which is a form of coercion, but which may be enacted without an actual rebellion or threat of secession. An example of this situation occurred during the Indonesian confrontation of Malaysia. The Malaysian Government and its allies believed that some Chinese youths living in the borderland were giving assistance to Indonesian irregulars, accordingly it was decided to concentrate these Chinese in controlled settlements at some distance from the border. This was a means of increasing state security which had a clear influence on the distribution of ethnic groups and settlement and farming patterns. An earlier example had occurred in Kenya during the Mau Mau rebellion, when the government reorganised the settlement of Kikuyu living at the foot of the Aberdare ranges, in order to deny rebels the chance of obtaining supplies and information without leaving the protection of the forest. Other instances of this situation, for example in South Africa, will be considered under the heading of administrative policies.

The third means of eliminating problems of dissidence involves the expulsion of the population concerned from the state. After the Israeli campaign in 1967, newspapers contained many references to allegations that the Israel authorities were expelling Arabs from the recently captured West Bank of the Jordan. If this is true it could be explained by the desire of the Israel Government to reduce administrative problems in the conquered area.

Conversely King Hussein of Jordan has urged Arab refugees to return to the West Bank area in order to make the situation as difficult as possible for the Israel administration. One of the most fully documented enforced migrations occurred in the Balkans between 1912 and 1925; according to Pallis (1925) there were seventeen migrations during this period. Ladas (1932) has produced an excellent account on which the following paragraphs are mainly based.

The first Balkan war began in October 1912 and this initiated various migrations. In a number of cases the civilian population followed victorious or defeated armies. For example, in 1913, during the second Balkan war, 15,000 Bulgarians joined the army in its retreat from Macedonia as the Greek forces advanced. But it is the planned movements, rather than spontaneous flight, that concerns the present account. There were four agreements governing the exchange of population in this area, two before the First World War and two after. In 1913 a Protocol, attached to the Turco-Bulgarian Convention establishing peace, dealt with the delimitation of the new boundary, and contained clauses for the exchange of the Bulgarian and Turkish populations.

> Les deux Gouvernements sont d'accord pour faciliter l'échange facultatif mutuel des populations bulgare et musulmane de part et d'autre ainsi que de leurs propriétés dans une zône de 15 kilomètres au plus, le long de toute la frontière commune. L'échange aura lieu par des villages entiers. L'échange des propriétés rurales et urbaines aura lieu sous les auspices des deux Gouvernements et avec la participation des anciens des villages à échanger.
> Des Commissions mixtes nominées par les deux Gouvernements procéderont à l'échange et à l'indemnisation, s'il y a lieu, de différences résultant de l'échange de biens entre villages et particuliers en question.
>
> (Ladas, 1932, pp. 18–19)

Ladas (1932, p. 19) notes that most of the people within the borderland had already crossed to avoid being in the position of an alien minority; the protocol therefore provided the machinery to compel the remainder to transfer and also to regulate the mutual compensation for farmlands and other immovable property abandoned by the refugees. The total movement involved 48,570 Turks and 46,764 Bulgarians, before the advent of the First World War halted operations. At the same time the

'Young Turks decided to implement their plan of ridding themselves of the national minorities and making the Turkish Empire a homogeneous state' (Ladas, 1932, p. 15). The Young Turks were particularly concerned about the security of the coast opposite the Aegean islands recently occupied by Greece. These coastal areas contained more than one million Greeks, and there was a fear that the irredentist spark might leap the narrow waters. Accordingly the Turkish Government began to deport Greeks from this area into remote Anatolia. At least 50,000 (Ladas also gives a figure of 85,000) were resettled in this way and a further 150,000 Greeks went voluntarily to Greece. These actions forced the Greek Government to agree to 'spontaneous' exchange of the Greek population of Thrace and the *vilayet* of Smyrna with the Turkish population of Macedonia and Epirus. The early entry of Turkey into the First World War prevented the application of this agreement. However, there had been sufficient time for 150,000 Turks to migrate back to Turkey from Macedonia.

During the war the Bulgarian army deported 36,000 Greeks to the interior of Bulgaria from the areas of eastern Macedonia which it occupied, and Turkey continued to transfer Greeks and Armenians from peripheral areas to Anatolia. After the war two agreements regulated the exchange of minorities between Greece and Bulgaria and between Greece and Turkey.

Greece and Bulgaria signed a Convention regarding reciprocal emigration in 1919, at Neuilly-sur-Seine. According to Ladas (1932, p. 9), Greek co-operation, at the request of the Great Powers, was designed by the Greek Government to 'preserve Hellenism in Turkey and to ensure the foundation of a real Great Greece'. This Convention was for a voluntary movement, but it did not prohibit regulations compelling persons to emigrate. There was a slightly greater willingness on the part of Greeks to move because they feared land appropriation under the social legislation programme of the Bulgarian Agrarian Government of Stambulinsky. A factor tending to reduce the movement of Bulgarians was the activity of the underground political party called the Macedonian Revolutionary Organisation. This group urged Bulgarians to stay and continue the efforts to establish Bulgarian ascendancy in Macedonia, which had showed some signs of success. After the first six months of operation by the Joint Commission, only 197 Greeks and 166 Bulgarians had indicated their intention to emigrate.

The situation changed as a result of conflict between Turkey and Greece. In 1923, to safeguard the strategic railway from Gumuljina to Dedeagach, Bulgarians were deported from its vicinity to Thessaly and certain islands. About 1,700 families were involved and this stimulated Bulgarian emigration; they were in any case beginning to face increasing economic competition from Greek refugees returning from Bulgaria and Turkey. This Bulgarian movement produced a Greek movement in reaction, because the Bulgarian refugees tended to batten on to Greek villages in Bulgaria and exhibit considerable hostility. The total migration after the Convention was implemented involved 53,000 Bulgarians and 30,000 Greeks.

At Lausanne, in 1923, Greece and Turkey signed a Convention for the compulsory exchange of their respective minorities. This Convention was part of a general peace agreed after the Greek military defeat in Asia Minor. The only Turkish nationals of the Greek Orthodox faith who would be exempted were the occupants of Constantinople, and the only Greek nationals of the Moslem faith to be exempted were those living in western Thrace. These exemptions were not clearly worded and there were disagreements over the meaning of the term 'established' and the definition of 'the prefecture of the city of Constantinople' as it was delimited in 1912. The significance of the Constantinople definition lay in the fact that it excluded the city of Pendik, which was in effect a contiguous suburb containing 15,000 Greeks of the Orthodox faith. The statistics of the Mixed Commission record that 189,916 Greeks transferred from Turkey and 355,635 Turks transferred from Greece in the period 1924–6.

In Greece, Turkey and Bulgaria the governments had to decide on policies regarding the resettlement of returned nationals. Turkey was not short of land and confined its aid to the establishment of Moslems on land vacated by Greeks, together with loans to purchase implements and seed. Greece faced greater difficulties because of the larger numbers of refugees, which included those who had fled before the Lausanne Convention (*c.* 1·069 millions), as well as those from Russia (58,000) and other countries (10,000), and the comparative shortage of land. To government land and land vacated by Moslems the government added 100,000 hectares secured by the compulsory acquisition of large estates. Bulgaria also faced serious problems, arising mainly from the apparently poor quality of much of the land, and had to request a loan from the League of Nations. This loan was granted, but a condition stipulated that as a rule refugees should not be resettled within fifty kilometres of the Bulgarian boundaries with Rumania, Jugoslavia and Greece. The Turco-Bulgarian borderland had already been settled in accordance with the 1913 agreement. This stipulation by the League sought to avoid potentially dangerous concentrations of people close to areas in another state which they had once owned and worked. The strict application would have eliminated 55 per cent of Bulgaria, and therefore exemptions were allowed where the boundary coincided with a major physical barrier. This allowed the settlement of 3,000 families along the Danube between Lom and Sistava. The river here was two kilometres in width and there were no bridges. The boundary between Greece and Bulgaria between the Struma and Moritza follows relief features associated with the Rhodope mountains which rise to 7,000 feet. Refugee families were allowed to settle in the Mastarla area, north of Kamotene, because of the major highland and the absence of significant trans-boundary roads, but a rider was added even here that no settlement should be within ten kilometres of the boundary, and any former occupants of Greece should be at least fifteen kilometres from the line.

Any survey of secessionist movements throughout the world since 1945 suggests a number of imperfect but useful generalisations. Some geographers, concerned to reduce information to indices, might try to grade the data about such movements, but they would defy their efforts. Factors such as remoteness, difficult terrain, degree of alien support and significance to the country concerned cannot be precisely measured, or if some measure is contrived, it cannot be compared with measurements of other situations.

The principal secessionist movements have occurred amongst sections of the Kurds in Irak, the Turkish Cypriots, the Mizos, the Somali of Kenya, the Tibetans, the Bantu peoples of the southern Sudan, the tribes of Katanga, the Ibo of the Eastern Region of Nigeria, the Baganda of Uganda and part of the Austrian population of the Italian Tyrol. These movements show wide political variations and some similarities in their degree of popular support, the vigour with which their cause is prosecuted, and their final aims. They also show some geographical similarities and differences. If these ten cases are considered it is clear that, with the exception of the Turkish Cypriots, the dissident groups are located in a compact area, and in six of the nine cases the terrain offers proven opportunities for defence and problems for attacking government forces—the jungles of the Mizo Hill Area and the ranges of Kurdistan are such areas. Again, with the exception of the Turkish Cypriots, and the Eastern Region of Nigeria, the regions involved are peripheral to the main state core. In all cases the dissident population is distinguished by one or more cultural features from the rest of the citizens. The differences between African tribes provides many obvious examples, as well as the religious differences between Turkish and Greek Cypriots and between Tibetans and Chinese, and the language distinctions between the Austrian and Italian inhabitants of the Italian Tyrol. In three cases, the Turkish Cypriots, the Kurds and the Somali, the dissident groups received aid and support from neighbouring states; and in five cases they had a history of either being independent or united with a neighbouring state with which they had clear cultural affinities. For example, the Tibetans had enjoyed complete autonomy during the period of internal Chinese weakness from 1913–1949, and Cyprus had been part of Turkey until 1878.

This brief survey suggests that secessionist movements are more likely to develop on the periphery of the state, especially in difficult or remote regions, when such an area forms a compact unit and is occupied by a group with distinctive cultural characteristics, who may be closely related to similar groups across the boundary, and who, perhaps, were once united with those groups or enjoyed a significant measure of autonomy. But, of course, the existence of all these characteristics will not produce a secessionist movement, unless some impetus is given, such as the rise of a popular leader or discrimination by the central government, or changed economic circumstances in the area.

Nor is it possible to predict the government's reaction to such movements. In Katanga, Tibet, Iraq, the Sudan, Uganda, Assam and Eastern Nigeria the governments have either used force or a combination of force and persuasion. In Cyprus the government is trying to use mediation. The situation in the Italian Tyrol is much less critical than elsewhere and normal police surveillance seems sufficient. The timing of government action will depend on a number of factors, amongst which may be included the importance of the area concerned both strategically and economically, and the international situation. India's evident concern with secessionist movements in the Himalayas must be a partial reflection of worsening relations with China since 1960. The Tibetan secessionist movement was relatively unimportant to

China until questions of leadership and survival against Japan had been resolved. The simple facts of climate may also exert an influence on the tactical timing of any operations against dissident groups in difficult terrain.

In the case of the Kurds, the Nagas and the Tibetans, the governments concerned have successfully enlisted the aid of part of the dissident cultural group. The aims of the secessionist movements may involve complete independence or unity with a neighbouring state, but the practical effect of success to the original state is the same—loss of territory. This is not a development which is easily accepted especially in those areas where the territory is strategically or economically important, or in the cases where one successful secession may lead to further fragmentation of the state. Secessionist movements are a source of weakness because of the diversion of military, administrative and financial resources, and because of the opportunity they afford hostile states to interfere; their removal will invariably strengthen the state. It can, of course, be argued that where the territory has no significant economic or strategic value, the state may well eliminate the problem by allowing the inhabitants to become independent or join another state. Policy decisions of this order are rare; territorial concessions to minorities are usually made after military defeat, as in the case of the Balkans, when the Turkish forces were beaten. It is also significant that the major periods of decolonisation, which is a form of secession, have occurred when the imperial states became weak or less strong.

It has already been indicated that secessionist movements provide an opportunity for hostile states to exploit the situation and foster the weakness of the state. This policy may sometimes be encouraged by the desire to annex the secessionist area.

> It is completely impossible for the Somali Government and the people of the Somali Republic to abandon the work they have been undertaking, which is that of liberating the Somali territories which are occupied forcibly by aliens, and of restoring their dignity and freedom to the inhabitants of these Somali lands.
>
> (Africa Research Bulletin, 1965, p. 396)

This view expressed by an information officer in November 1965 has subsequently been borne out by the actions of the Somali Government towards Somali minorities abroad. These actions have allowed free movement across the Kenyan and Ethiopian boundaries, providing a sanctuary for rebels in time of need. In the case of the French Territory of the Afars and Issas, during the period leading to the referendum on the territories' future status in March 1967, the Somali Government was accused by the French of flooding Djibouti with illegal Somali immigrants to swell the vote in favour of independence. Six thousand Somali were deported before the referendum, but then the Somali Government refused to accept any more deportees, presumably in an effort to counter French efforts to remove the problems of dissidence by evacuation.

Reference has already been made to a similar tactic by the Government of Jordan which has urged refugees from the war with Israel to remain on the west bank of the Jordan river, so as to create problems for the Israeli authorities, and reduce the Jordanian burden of feeding refugees without access to the most fertile part of the country.

The efforts of Russia to exploit the Kurdish situation in Iran have been described by Hassan (1966).

> It appears that Soviet policy towards the Kurdish question had not taken definite shape at that time [the joint Anglo-Soviet occupation of the area] since there were probably conflicting views on the problem. From the point of view of general politics the Soviet Government would have liked to help in the creation of a Kurdish State, comprising the Kurds of Iran, Turkey and Iraq. In this way the anti-communist and anti-Soviet regimes of Turkey and Iraq, which were considered to be under Western influence, would be considerably weakened, and a door opened to Soviet infiltration to the Middle East. On the other hand, such a state covering important regions not occupied by the Soviet Union might show a spirit of independence which would not coincide with Soviet policies.
>
> (p. 71)

This was the Soviet dilemma and they decided to encourage the secession of the area to join with the Azerbaijan S.S.R. Use was made of a group of Kurds from the Ahar, Sarab and Ardabil districts of Iran. These people had been crossing into Russia since the discovery of oil at Baku and the establishment of the port, to find work. Many returned when they had earned enough money, but others stayed and married Azeri women. In the thirties 50,000 of these people were expelled from Russia and resettled on the orders of the Shah in their former homelands. However, by this time they had become urban-dwellers and they congregated at the town of Ardabil. This group became discontented with economic conditions, and in 1945 it was armed by the Russians, and allowed to revolt at Mianeh, then Tabriz and Ardabil. An autonomous Republic was formed on 15 December 1945. The leaders of the new state showed little enthusiasm for union with the Azerbaijan S.S.R. and friction with that state developed over the ownership of the towns of Khoi, Rezaiyeh and Shahpur, and the fertile plains surrounding them. An agreement was signed in April 1946, which included the guarantee that any negotiations with the Iranian Government would be carried out with the full knowledge of both parties.

Russia withdrew its forces in the following month, partly due to pressure from the Security Council, but mainly because of the desire to negotiate an oil concession in northern Iran. The rebellion then collapsed within ten months.

Before leaving this question of dissidence it should be noted that governments may also adopt policies which will prevent the formation of significant minorities. Rose (1966, pp. 47 and 84–7) attributes the Australian immigration restrictions in respect of Asians to the fear that excessive immigration would create a fifth column.

The mobilisation of material resources

A second means by which a government may make the state stronger is through the more complete mobilisation of the country's material resources. This will involve increased production, use of available substitutes, the securing of guaranteed supplies from other states, stockpiling and many other actions. Conversely a state may seek to weaken a hostile state by denying it access to essential raw materials and other economic processes.

It is generally accepted now that if a country is involved in a war the government has the right and the duty to control all productive capacity in such a way that the country's success is secured in the shortest possible time. In non-totalitarian states this requires a much greater involvement of government in planning and co-ordination than during peacetime and the geographical consequences of this are correspondingly greater. A prime aim of governments at such times is to increase self-sufficiency in ways which would probably be uneconomic during peacetime. This is in response to the need to produce as many of the country's needs as possible, in the territory under the government's control, free from the risk that outside supplies will be withheld, or prevented from arriving by blockade.

Food production is an immediate target for improvement, and one of the best examples of this subject is to be found in the description of Britain's agricultural policy during the Second World War by Murray (1955). This is a volume in the Civil Histories series, which has proved most useful in the preparation of this material. The various volumes on production, manpower, agriculture, the economy and trade are based on primary government sources, although these are not yet available to scholars. This use of such detailed material gives a real insight into the relations of geography and these aspects of policy.

It had been decided in 1926

> that no case had been made out on defence grounds which would justify the expenditure necessary to induce farmers in time of peace to produce more than economic conditions dictated.
>
> (Cmd 2581, 1926)

This view was apparently based on the belief that any advantages in national security would be more than outweighed by alienation of the sympathy of overseas countries, including members of the Commonwealth which found their markets curtailed; a reduction in the amount of shipping at Britain's disposal and a curtailment of shipbuilding. But this policy began to be questioned in the spring of 1935 when Germany openly breached the Treaty of Versailles, and a committee was appointed to make recommendations about the production of food in wartime. The aims postulated by this committee a year later were an immediate increase in agricultural activities which would yield the greatest increase in food value and foodstuffs which were bulky to import, special efforts to improve the production of wheat, potatoes, sugar beet, oats and eggs, and the maintenance of pre-war levels of animal products so far as possible. Sugar beet, potatoes and wheat yield food value in that order, but when this is related to labour involvement, which is clearly critical, wheat becomes most important. The targets identified by the committee were not entirely compatible; for example, wheat production could only be increased by ploughing up grassland, but this would reduce the output of animal products unless foodstuffs could also be grown or imported. The committee also indicated that these aims would require government action to provide sufficient machinery, labour and fertilisers.

In 1936 Germany's reoccupation of the Rhineland, and further developments which seemed to make war more certain, encouraged the government to carry planning a stage further. A new committee was instructed to assume that if there was a war, there would be a severe temporary dislocation of overseas supplies for a period of three months, that food imports would only amount to three-quarters of the peacetime level, and that while the Mediterranean would remain open the Baltic would be almost totally closed, and that the North Sea route would be impaired. The committee recommended first a storage programme to overcome temporary dislocation (Hammond, 1951, vol. I, chapter II), and second a programme designed to increase agricultural output. The storage programme concentrated on livestock feed and fertilisers, specifically phosphates, sulphur and ammoniacal liquor. This programme was eventually judged too costly, with the consequence that little was done, and Britain started the war with negligible amounts of feedingstuffs and fertilisers.

As the situation in Europe became more ominous, attention was focussed on improving soil fertility, as an alternative to the storage programme, so that if war was declared, production could be quickly expanded. This aim was partially secured by the government subsidising the use of lime and basic slag, and by the modification of the 1930 Land Drainage Act, to encourage drainage of potentially useful areas. In April 1939 a new Minister of Agriculture sought to accelerate preparations for wartime production by requesting and receiving permission to purchase additional stocks of phosphates, 5,000 tractors and to subsidise arterial drainage and the conversion of permanent grasslands to arable land. The rate offered was £2 per acre and it applied throughout the war.

During the war a series of production targets were defined. In 1940 and 1942 the desired annual increase in the area of arable land was two million acres; in 1940 the total achieved was 1·6 million acres and in 1942 1·1 million acres. It was clear by 1944 that this programme had just about achieved its maximum extent. In that year 600,000 acres of grassland were ploughed and 506,000 acres of arable land reverted to temporary grass. Not all the targets were achieved, and it became evident that certain factors which had not been considered were important, and that certain premises had proved false. For example, the assumption that food imports would run at three-quarters of peacetime levels was hopelessly optimistic, as the following figures of imports by the Ministry of Food show.

	Millions of tons
Pre-war	22·5
1939–40	20·7
1940–1	14·4
1941–2	12·7
1942–3	10·3
1943–4	11·5
1944–5	11·6

(Source: *Statistical Digest of the War*, H.M.S.O., 1951, Table 161)

Delays in ploughing were caused by the fact that machinery was principally located in the eastern part of England, while the largest available areas of grassland were in the west. The lack of arable farming experience in certain predominantly livestock areas was underestimated as a significant factor. There was a much more serious drain of agricultural labour to the forces and other occupations before special legislation was introduced in 1940. To these wrong guesses can be added such uncontrollable factors as the three severe winters and the dry springs of 1939–41 which reduced milk production.

The application of these wartime policies had a clear geographical influence on the landscape and the farming industry. According to Murray:

> . . . none of the 355,000 regular farmers in the United Kingdom and few if any of the other 190,000 farmers to whom farming was not a full time occupation can have carried on with their pre-war system of farming or persisted in their habitual farming methods. Each must have adapted his farm in his individual way, to meet the new circumstances dictated by the war.
>
> (1955, p. 249)

A major change was the conversion of 6·3 million acres of permanent and temporary grasslands to arable production between 1939 and 1943. The effects of this campaign were felt after the war when 'ley farming', which had increased during the war, was continued despite a reduction in the labour force. Ley farming allowed higher productivity from temporary grasslands and maintained a higher level of humus in the soil, which was better textured. This reduction in the area of grassland was not matched by a proportional decline in the livestock levels; and this did not merely represent taking up the slack which had developed during the Depression, it involved a real increase in the carrying capacity of the pasture. The war also introduced greater use of the Bersee and Atlé varieties of wheat, which yielded so well in some areas that their use continued after the war. Fertilisers were used on an increasing scale during the war. In 1939 farmers used 77,100 tons of nitrogenous fertilisers, 195,500 tons of phosphates and 85,000 tons of potash; by 1943 the comparable amounts used were 181,500 tons, 343,600 tons and 103,700 tons. This development would probably have occurred in any event, but the immediate needs of the war accelerated the process, just as it accelerated the introduction of machinery to replace scarce labour. Between 1942 and 1946 the number of tractors increased from 166,830 to 203,420, combined drills increased from 7,930 to 17,040, hay and straw balers increased from 2,760 to 6,680, combine harvesters from 1,000 to 3,460 and milking machines from 29,570 to 48,280. This mechanisation continued after the war as the drift of labour to the towns continued.

But perhaps the most important result of the war was the decision of the National Government in 1940, recognising 'the importance of maintaining, after the war, a healthy and well balanced agriculture as an essential and permanent feature of national policy' (House of Commons Debates, vol. 367, Col. 92). Murray has recorded that the foundations of the post-war policy were decided as follows, in July 1942:

(1) that all reasonably good agricultural land should be maintained in a state of productivity and fertility, and
(2) that any policy must secure to the utmost practical extent proper standards of upkeep of the farm and farm buildings, proper standards of farming and economic stability for the industry.

(p. 348)

The acceptance of these dicta was responsible, in the post-war years, for the charges of feather-bedding of the farmers, as peace-time economic conditions came into conflict with the government's concept of the farmer's contribution to national security.

If a country is to defend itself it needs certain raw materials as well as food; materials such as iron ore, petroleum, flax and mineral fertilisers. Such resources will usually be sought in a combination of three methods: increased home production where possible, the substitution of available alternatives, and agreements to guarantee supplies from other countries. Such diverse activities clearly require a measure of control and co-ordination; this is generally exercised by the government, and provides further evidence that government involvement in the planning increases during a war and may be continued at a higher level than normal after the war. The success of these policies will be directly related to a number of non-geographic and geographic factors. Into the first group will fall the skill and foresight of the planners and the ingenuity of scientists. The main geographic factors will include the quality and quantity of home resources, location of supplies in friendly and neutral countries, the nature of the routes linking the resources and the state, and the quality of the geographic information available to the planners. Jones (1954) has provided an excellent account of how resources should be evaluated in terms of the time which must elapse before they can be made available. To this may be added their defensibility, since resources in exposed positions, liable to air damage or capture, may be worth less than poorer resources in more sheltered locations.

Hurstfield (1953) provides a good example of different assumptions by policy-makers in the case of Britain and Germany in the thirties.

> Great Britain could, on the whole, feel legitimately secure with her existing raw material supplies, which she could only lose as the result of major defeats in theatres of operations affecting the chief sources, or through heavy losses at sea. Neither situation was seriously envisaged, as is reflected in the absence of large raw material stocks and of schemes to develop alternative sources at home.
>
> . . . Germany knew that in the outbreak of war with the Western Powers she would lose the large bulk of her sea-borne supplies. On this assumption she adopted the two-fold policy of building up close relations with her neighbouring states, which in due course led to full control, and secondly of establishing an important series of industries for the production of substitutes.
>
> (p. 30)

The development of stockpiles is now common practice for major powers. In addition America has imported certain commodities, mainly petroleum, of which she has adequate amounts, to conserve domestic supplies. Policies of this nature have a significant influence on questions of the profitability of production, the areas of output, and the patterns of trade which ensue. Strategic planning may also encourage the search for important resources as well as their stockpiling. The best current example is provided by the South African Government, which is stimulating an extensive search for petroleum in western and off-shore areas.

Hurstfield's account of the raw material situation in Britain during the Second World War is of the same high standard as the other volumes in the Civil Histories series, and furnishes many interesting examples of how Britain attempted to increase its scale of domestic production. For example, the output of hardwoods and softwoods increased from 444,000 tons and 496,000 tons respectively in 1940, to 720,000 tons and 1,129,000 tons respectively in 1943. This increased scale of production meant that there was less reliance on imported woods, which in 1943 only supplied 42 per cent of the country's needs, compared with 77 per cent in 1939.

There was a similar decline in the imports of iron ore. In 1939 imported ores represented 25 per cent of the total used, but since the British ores were leaner, this represented 50 per cent of the metallic content available. By 1942 imported ores only represented 9 per cent of the total used by weight. The change resulted from the increased production of British phosphoric ores. These ores were suitable only for the manufacture of basic steel which was regarded less favourably than acid steel for certain engineering processes. The haematite ores used in the manufacture of acid steel could only be obtained from a few locations in Britain, and the Cumberland field, the major source, was facing severe problems of subsidence and shaft-sinking, which resulted in a decline from 761,000 tons in 1939 to 535,000 tons in 1944. The switch from the use of acid to basic steel was not the only problem, because many steel works had been built close to ports, such as Glasgow, Cardiff and Newcastle, through which most of their overseas supplies arrived. These plants were now supplied from Northamptonshire and Oxfordshire, increasing the strain on internal transport.

The efforts in respect of iron ore were more successful than in the case of the majority of base metals, for in 1942 the Ministry of Supply decided to defer any further attempts to increase output of tin, wolfram and zinc. Bauxite deposits in Northern Ireland, although possessing only half the metal content of imported ores, were developed and reached a peak output of 90,000 tons in 1943; increasingly, however, Britain relied on the import of alumina from Canada, which saved valuable shipping space.

To summarise the situation, efforts to make a country more self-sufficient will normally be taken at times when the country faces a threat of war, or when it is planning to start a war, or after a war has started. These efforts will be directed by the government which will often assume complete control over the development and distribution of resources. The prime aims are to reduce dependence on uncertain sources of supply, and to reduce strain on external transport. The geographic effects of such policies will vary. These policies may result in the rejuvenation of areas which had declined during the operation of peace time economic conditions, or result in the rapid development and improvement of transport links to serve the increasing volume of internal traffic. The most obvious effects in agriculture will probably centre on an intensification in the production of industrial and food crops at the expense of fallow land and permanent pasture, and an extensive clearing of forest areas to replace reduced bulky wood imports. The effects are likely to continue into the post-war period because governments are often slow to dismantle control-machinery and abandon powers which have been assumed, and because the emergency conditions of war will often telescope changes in agriculture and industry which would have taken much longer under peaceful conditions. This was seen in Britain, not only through the increased use of fertilisers and farm machinery, but also through the development of plastics and artificial fibre industries. Further, the policy of encouraging substitution will frequently show an effect in post-war conditions by depressing raw material prices; as the rubber and dairy industries have found.

But self-sufficiency and substitution are only two of the avenues of government policy—the third concerns the use of raw materials from other countries or territories. Such materials may be purchased, or secured by conquest; in either case geographic factors will be vital. Looking first at the case where the resources are obtained by purchase a number of factors will be involved, which will make different policies more appropriate at different times.

Distance and location will normally be the most important factors. Nearer resources will usually be preferred providing the routes are safe, even if the closer deposits are of inferior quality. The overseas territories from which supplies can be obtained fall into three groups. First, there are the colonies or overseas territories which can be regarded as an extension of the domestic resource base. Britain, France and the Netherlands were fortunate in their possession of African colonies during the Second World War; by contrast Italy was poorly served by Libya, Somalia and Ethiopia, and Germany's colonies had been lost twenty-four years earlier. In such areas the government can exercise authority and there are no currency problems, which can be a significant advantage. The second group is composed of allies or sympathetic states, which could be expected to give favourable terms and as much assistance as possible. Third, there are the neutral states. Such states provide resources to both sides, and for this reason they are especially important. It is a frequent policy to derive as many supplies as possible from neutral states in order to deprive the enemy of them. This situation offers a considerable opportunity to the neutral states, since prices tend to rise, and by barter the neutral states can often obtain scarce items which they need, but which they do not possess. Policies dealing with the provision of overseas supplies need to be flexible because of the major changes which are possible as a result of military and diplomatic action. Sometimes the changes will be geographic in nature: the loss of territory or the discovery of new mineral deposits or the development of new plant strains. Sometimes they will be non-geographic: currency difficulties, a strike by workers in a critical industry or the loss of a significant tonnage of shipping. But whatever the cause, the effect is likely to be new patterns of trade and distribution.

Hurstfield's account of the changes in British sources of overseas supplies could serve as a model for political geographers interested in this field. The following account is based partly on Hurstfield's analysis and partly on an equally fascinating study by Medlicott (1959) dealing with the economic blockade of Germany.

In 1938 Britain drew imports from three main areas: Europe, the Commonwealth countries and North and South America. Europe supplied most of the iron ore (55 per cent), steel (57 per cent), bauxite (91 per cent), magnesite (51 per cent), ferro-alloys (85 per cent), iron pyrites (91 per cent), flax (85 per cent) and softwoods (56 per cent). Commonwealth countries provided most of the chromium (90 per cent), manganese (98 per cent), tin metal

(72 per cent), tungsten (82 per cent), zinc ore (69 per cent), lead (61 per cent), wool (82 per cent), silk (86 per cent), hemp (soft 51 per cent, hard 100 per cent), jute (99 per cent) and rubber (99 per cent). The American continents supplied the bulk of Britain's supplies of antimony (76 per cent), aluminum (62 per cent), wrought copper (72 per cent), molybdenum (97 per cent), tin concentrates (68 per cent), wrought zinc (58 per cent), nickel (100 per cent), cotton (54 per cent) and hardwoods (52 per cent).

The outbreak of war deprived Britain of important sources of potash, and magnesium from Germany, timber from Poland, and magnesite from Austria. German control of the Baltic also prevented British access to its usual sources of Russian flax and Swedish softwoods. To make up these losses, potash was increasingly obtained from Palestine, Greece doubled its output of magnesite, magnesium and softwoods were obtained from American sources when shipping space allowed, and flax production in Northern Ireland was expanded. The acreage rose from 25,700 in 1939 to 147,600 in 1942 and 198,400 in 1944. The German advance into Norway severed connections with Swedish iron ore, which was of very high quality and used in the acid process of steel-making, and with the remaining supplies of Swedish softwoods and with Norwegian supplies of pyrites, which was used in the manufacture of sulphuric acid. Iron-ore supplies at this time were still available in France, Spain and North Africa, while pyrites could be imported from Spain, Portugal and Cyprus.

The next German advance into the Low Countries and France, and Italian successes in Greece, worsened the supply situation. Supplies from the French ore fields were lost and supplies from Bilboa, on the north coast of Spain, and from North Africa were placed in danger from air attack. Bauxite from France and Greece was no longer available and Greek magnesite was lost. Finally, supplies of flax from Belgium and phosphates from North Africa were halted. The iron-ore deficiencies were made up by increased imports from Brazil (31,000 tons in 1939; 261,000 tons in 1943), Sierra Leone (190,000 tons in 1939; 920,000 tons in 1941), and South Africa (2,000 tons in 1939; 247,000 tons in 1941). Bauxite was increasingly produced from Northern Ireland's low-grade deposits and the balance was made up by the import of virgin metal from Canada and by an intensive salvage campaign in Britain. Magnesite was derived from Australia, Russia,

South Africa and the United States, while phosphates were brought from Palestine and Florida.

It was unfortunate that some of the most important items lost through Axis conquests were also the bulkiest, such as iron ore and timber, which placed a much greater strain on British shipping than the greater percentage loss of ferro-alloys. Hurstfield (1953) stresses this point:

> Indeed, it was in its effects upon shipping, and to a lesser extent, upon currency, that the German conquest of Europe had its most significant consequences . . . As long as exchange difficulties made increased purchases from the dollar area undesirable, these alternative supplies were bought as far as possible in Europe and other relatively soft currency countries.
>
> (p. 159)

The other dramatic impact on patterns of supply was the Japanese advance in the Pacific in 1942. This resulted in a loss of major sources of rubber, tin, hemp, tungsten and silk, and minor sources of lead and antimony ore. Rubber and hemp losses were most serious because of the impossibility of rapidly bringing new areas into production, or expanding existing areas of output. Ceylon was the only remaining large-scale rubber producer and by 1943 production was only 6,000 tons above the 1941 level of 99,500 tons; but this total proved to be a peak, for adverse weather, shortages of labour and over-tapping caused a decline in subsequent years. Attempts to improve production in West and East Africa and Ethiopia were unsuccessful in meeting the need, which was increasingly satisfied from synthetic sources. Hemp could be substituted by sisal, but the industry in East Africa had been allowed to run down in response to overproduction and falling prices, and the 1941–4 period showed only a 30 per cent increase in output. Tin needs could be met more easily by increased imports from Nigeria and Bolivia, while silk was replaced by artificial fibres such as nylon and rayon.

This suggests a general conclusion that mineral losses can often be more easily replaced because of the possibility of rapidly increasing output, whereas in the case of tree-crops, such as timber, cocoa, rubber and palm oil, it is much more difficult to rapidly expand output, and this will frequently encourage a search for substitutes.

During this period Britain's supplies from the continent did not stop completely, for there were six neutral states, Sweden,

Luxemburg, Spain, Switzerland, Portugal and Turkey. An examination of Britain's policy towards these states shows the influence of various geographical factors. The location of the states was of prime importance; Luxemburg and Switzerland were surrounded by Axis-controlled territory and this reduced the pressure which Britain could bring to bear, and the ability of these states to trade with Britain. Sweden was effectively surrounded by German land and sea forces, which reduced trade possibilities with the rest of the world. These three states were all more effectively within the German sphere of influence than Portugal and Spain, which had a common land border, and Turkey, which was quite detached from Axis territory. The location was also important in terms of future plans. Spain and Portugal, and to some extent Vichy French North African areas, were more sensitive because of the proposed invasion of North Africa. The third aspect of location which was important was the ability of the neutrals to play the role of middleman between Germany and other overseas countries, especially in South America. Clearly the Iberian countries had greater scope for such action than Turkey. An example of this situation is provided by the fact that Spain was an important link in the Linee Aeree Trans-continentali, which flew Rome, Spain, Dakar, Brazil, and which was extensively used to smuggle small-volume, high-value goods, such as mica, diamonds and platinum. This line was stopped in 1942 when pressure on the Standard Oil Company of New Jersey ended sales of aviation fuel in Brazil. Thereafter this contraband travelled by ships via the Canary Islands. The nature of the resources available in the neutral countries, and the neutral needs which the Allies could satisfy, were also important factors influencing policy. The main items exported to Germany were iron ore and ball bearings from Sweden, ammunition from Switzerland, chrome from Turkey and wolfram from Spain and Portugal. In the first two cases Britain was concerned to restrict these supplies to Germany, whereas attempts were made to secure supplies of chrome and wolfram from the other countries. The nature of the economy of the neutral state was also important, especially in the case of Spain. This state had suffered severe economic dislocation during the civil war, and unless supplies were provided 'at least in some reasonable measure from overseas, the Spanish Government might be driven into full collaboration with the Axis' (Medlicott, 1959, p. 284). It is also apparent that historical association is a significant factor. Britain sought much more direct pressure against the South American states which still sent supplies to the Axis, while the United States urged caution. The situation was reversed in Europe with Britain resisting American attempts to coerce Sweden and Portugal.

In each case the policy adopted towards a particular neutral was finely balanced to ensure that assistance to the Axis was kept above the minimum which would *provoke* Axis occupation, and that supplies provided to the neutrals, by the Allies, was below the level at which stocks could be accumulated, which would *attract* Axis occupation.

Of course, this situation offered considerable opportunities to the neutral countries as well as problems. The prices of raw materials tended to rise, sometimes quite dramatically. In 1941, Britain was buying Turkish chrome at $22 per ton, in early 1942 this had risen to $27, and later in the year Germany was paying $30 while the price to Britain was $28. In 1943 Germany agreed to a price of $54 per ton. Wolfram prices showed spectacular growth. Spanish wolfram in February 1941 cost £675 per ton; thirteen months later it had reached £4,063 per ton, while Portuguese wolfram had risen in price to £6,000 per ton. This increase resulted in the abandonment of farming and some urban jobs in the wolfram areas, north of the Sierra da Estrala. These workers started to fossick for wolfram, and they were following the lead of miners at the British-owned Beralt mines, who created serious labour shortages when they left to discover their own supplies of wolfram. As this development coincided with a Portuguese Government campaign to grow as much food as possible, it is not surprising that the government introduced a measure of control in 1942 to drive people back into farming.

In similar fashion South American exporters of minerals had been making high profits through the sale of tungsten (Bolivia), copper (Chile), mercury (Mexico) and beryl (Brazil) to Japan, and Swiss trade with Germany was very profitable.

Before examining the alternative method of obtaining overseas supplies by conquest, it seems worth while to study the converse of these policies to strengthen the state by access to raw materials, which is the attempt to weaken the enemy by denying them access to vital supplies.

Blockades are a common feature of history; a good example is provided by the British blockade of Dutch ports from 30 November 1832 until 1 May 1833 as part of Anglo-French efforts to coerce Holland into accepting the twenty-four articles. For a period it was sufficient to besiege the town or castle, which was the strongpoint of resistance; subsequently, with the improvements in communications and the exercise of political control, it became necessary to treat the whole country, or large parts of it, as a unit of production, and blockade the whole state. The French blockade of Britain, instituted by the Berlin decree of November 1806, and the British reaction through Orders in Council three months later, were good examples of the early application of this concept of weakening the enemy. It is interesting in this case that Napoleon thought that the main aim of the blockade was to prevent the export of British goods.

There is a modern parallel to this view in the determination of the United Nations to block Rhodesian exports. This policy of not worrying about imports will only be successful if the matter is being fought only with economic weapons, and providing the country cannot revert to a condition of subsistence without serious internal disturbances. Although the methods of control have become more sophisticated the aims are still the same and there are still two areas of operation. First there is the possibility of controlling the sources of the contraband, second there is the chance of controlling the lines of communication by which it reaches the enemy. Both these means may be used against supplies from neutral countries, whereas only interruption to communications can be employed against supplies being moved from one enemy area to another. In the Second World War Japan and Germany exchanged some supplies via Russia's trans-Siberian railway, a route which was free from the chance of interference. In 1940–1 it is known that the Russian Government arranged for the movement of 600,000 tons a year from Japan to Germany. This safe route was severed in June 1941 by Germany's attack on Russia, and left the voyages around Cape Horn and the Cape of Good Hope as the only alternatives. These distances were 17,000 miles via Cape Horn, without the possibility of refuelling, and 11,000 miles via the Cape of Good Hope with the possibility of refuelling at Madagascar. These routes were constantly patrolled to prevent imports by Germany, and Britain made it a priority to prevent any link between Japanese and German forces in the Middle East. This was the prime aim listed by Lord Selbourne, Minister of Economic Warfare, when he assumed office in 1942 (Medlicott, 1959, p. 15).

The actions to gain control of production in a neutral country vary with the countries concerned. In some cases it may be that the state owns some of the units of production in the neutral country, such as mines, farms and factories. This was certainly the case in the Portuguese wolfram industry, where both Britain and Germany had acquired control of certain mines. A converse of this situation has been suggested by Rose (1966, p. 91) in connection with the refusal of the Australian Government to accept certain investments from Asian countries. In other cases the state may seek to purchase all the output of a particular commodity, whether it needs it or not; this policy is usually described as pre-emption. It is only possible to employ this method in cases where the commodity meets certain conditions. The commodity must be vital to the enemy's war effort; its production must be strongly localised so that alternative sources cannot be readily tapped; it must be bulky, otherwise efforts may be thwarted by smuggling, for example it would be pointless to pre-empt diamonds or mica; and finally it should be incapable of artificial production. Pre-emption will also be most successful where the neutral country is unable to expand rapidly production of the particular item, otherwise there is a danger that higher pre-emptive prices may stimulate production, making it necessary to spend more money in denying the surplus to the enemy, although it may be an advantage to raise the price if the enemy is short of foreign exchange. In the Second World War British and American pre-emptive policies were mainly employed in the Mediterranean countries, partly to prevent Germany obtaining supplies, and partly to drive the price up and force Germany to exhaust her foreign exchange holdings.

A third method lies in persuading the neutral state to impose its own controls. This can be done through a threat to supplies needed by the neutral and controlled by the state making the request, or by diplomatic pressure. In July 1941, when the Swiss-German trade agreement was concluded, Britain stopped all exports to Switzerland other than food and oil for soap-making. Sweden was also subject to pressure, in respect of oil supplies, to make her reduce exports of ball-bearings to Germany.

The second theatre of policy relates to the routes along which supplies are transported to the enemy. Control tends to be exercised close to the origin or destination, or at points, such as the Cape of Good Hope, and the straits between Sicily and Tunisia, where shipping channels are restricted in area. One of Britain's fears over the German conquest of Norway was that it would give German vessels access via Norwegian territorial waters, known as the Leads, to the open sea where they would be difficult to control.

> . . . The existence of the Leads . . . enabled German ships to enter territorial waters at remote points well inside the Arctic Circle and travel under their protection almost as far as the Skagerrak, where the proximity of German air and submarine bases made the rest of the journey comparatively safe from British interception.
>
> (Derry, 1952, pp. 9–10)

The fourth method was by blacklisting certain firms which were known to be dealing with the enemy. This was a technique which was used by Britain and America against South American and Swiss firms. Blacklisting has also been used by both the Arab states and Israel, since 1948, in an effort to dissuade the firms from establishing branches in the other territory.

Africa has provided two recent examples of economic sanctions, first, when Rhodesia declared itself independent in November 1965, and second, when the Eastern Region of Nigeria declared itself independent as the Republic of Biafra in May 1967. In each case the countries were subject to immediate economic sanctions. For example, on the day that Rhodesian independence was declared, the British Government announced that all British aid to Rhodesia would cease, that exchange controls would apply, that the export of British capital to Rhodesia would be banned, that Rhodesia would be denied access to the London capital market, that the Ottawa Agreement (1932) which governs trading between the two countries would be suspended, that Rhodesia was excluded from the Commonwealth Preference Area, that imports of Rhodesian tobacco were prohibited, and that the Commonwealth Sugar Agreements in respect of Rhodesia were abrogated. On 6 December 1965 other fiscal measures were introduced and imports of asbestos, copper, citrus fruits and juices, iron and steel products, maize, meat, chromium ore, antimony ore, lithium ore, tantalum ore, tea and dried vegetables were prohibited from Rhodesia. On 17 December 1965 two British orders prohibited the imports of oil and oil products into Rhodesia and the carrying of such products by British subjects for Rhodesia. In April 1966 the British Government, with the approval of the United Nations, enforced a naval blockade to prevent oil for Rhodesia being discharged at Beira. In November 1966, the Security Council of the United Nations instructed all members of the organisation to prohibit the imports of twelve items from Rhodesia: asbestos, chrome, iron ore, pig iron, sugar, tobacco, copper, meat and meat products, hides, skins and leather. These actions by Britain and various other states which supported her resulted in defensive reactions by the Rhodesian Government. Gradually, financial controls were established which prevented the export of dividends, rents and interest, and limited the amounts of capital which could be transferred from Rhodesia, international debts guaranteed by Britain were repudiated, petrol rationing was introduced, restrictions on the employment of alien African labour and the retrenchment of employees were implemented, the tobacco farmers were guaranteed a minimum price of 26d per pound in 1966, in return for which the government assumed the right to set a limit on the area which may be planted to the crop. In July 1966 the target for the 1967 crop was fixed at 200 million pounds at a guaranteed price of 28d per pound. This was a decline of 50 million pounds on the 1966 crop. The target for the 1968 crop was set at 132 million pounds in June 1967, and at the same time it was announced that tobacco farmers could sell all or part of their quota at the rate of 6d per pound, in order to raise funds to help in the conversion of their farms to other crops. These actions by the opponents of Rhodesia's government, and the reactions of that government, had a number of evident geographical effects.

First there are the changes in the pattern of international trade, changes in both direction and volume. The diminution of trade between Rhodesia and Britain has been partly balanced by the increased trade between Rhodesia and South Africa and Mozambique, and by increased trade between Britain and various areas of tobacco and mineral production in the Americas. Within Mozambique, the oil pipeline, from Beira to Umtali, has been replaced as the major avenue of oil import to Rhodesia, by the railway from Lourenço Marques. Zambia, which formerly conducted most of its trade through Rhodesia and Mozambique, still exports most of its copper via this route, but receives most supplies, including petroleum, via air and land routes from Dar-es-Salaam. Whittington (1967) has examined the effects of the Rhodesian crisis on Zambia. The Benguela railway has also been carrying a higher volume of Zambian imports and exports. Second, there are the internal changes in Rhodesia, which include the increased mining of gold, the diversification of tobacco farms, and the establishment of local industries to provide substitutes for unavailable imports. There has also been a slackening in the flow of migrant workers from Zambia, seeking employment in Rhodesian mines and on Rhodesian farms. Stock piles of Rhodesian exports have been accumulated in Rhodesia and Mozambique, against the day when prices may start to rise again. For example, it was announced in June 1967 that by the end of the tobacco sales in 1967, Rhodesia would have a stockpile of 210 million pounds, or more than the production total in either of the previous two years. It is possible that some of these temporary changes may become permanent, and that more significant changes will occur, such as the completion of a railway northwards from Beit Bridge or closer economic union of Rhodesia and South Africa.

The sanctions applied against Biafra by the Federal Government of Nigeria were also declared on the same day that independence was assumed—30 May 1967. According to the regulations, the exchange of Nigerian currency with the Eastern Region was prohibited, and so were foreign exchange transactions; bank balances held by residents in the Eastern Region, in other parts of the country were frozen; and all postal and telephone communications were severed. In addition, the Federal Government established a prohibited zone, which was defined as being

> east of a line Cape Formosa and position 04°04′ North and 06°05·5′ east and a line parallel at a distance of 12 miles from the coast of Nigeria until the international waters of the Republic of Cameroun are reached off the Calabar river.
>
> (*Nigerian Evening Post*, 2 June 1967)

All ships, including fishing craft, were prohibited from entering this area. Oil tankers taking oil from the Mid-West State to Port Harcourt, or oil and refined products from Bonny and Port Harcourt, were required to seek prior permission from Lagos. This permission was generally granted until early July, when oil traffic was blocked because of alleged payments made by the major oil company to Biafran authorities against the instructions of the Federal Government. In addition to these measures, the Cameroun boundary with the Eastern Region was closed by the Cameroun authorities, and negotiations were reported to block the traditional smugglers' route from Calabar to Fernando Po. Shortly after the cessation of oil exports the Federal Government sought a military solution to the secession.

This brief review of blockades during the Second World War against the Axis, and as a means of quelling rebellion in Africa, together with the useful review of the possibilities of sanctions against South Africa, edited by Segal (1964), and the various studies of the abortive sanctions against Italy in 1936, suggest a number of general geographical conclusions.

First, the location and shape of the country will have a significant influence on the logistic problems of enforcing any blockade. Location will determine whether the state is an island, in which case the blockading state may be able to act effectively alone, as the American blockade of Cuba revealed. There are clearly a variety of cases, however, which would pose different problems, the differences between very large islands and small islands, the cases of islands, such as Greenland, where much of the coast is inhospitable for part of the year, the cases of islands which are so close to the mainland that there is no area of high seas between. The state being blockaded may be marginally placed, in which case the number of adjacent states will be significant, for their co-operation will usually be required. It will normally be easier to blockade a marginal state, where the aggressive power is the only land neighbour of that state; examples of this situation are provided by the Gambia and Senegal, Hong Kong and China and Portugal and Spain. Marginal states which possess access to two or more oceans, such as France, Mexico, Russia and South Africa, will present greater problems. An exception to this rule is Israel, which has access to the Red Sea and the Mediterranean; the access to the Red Sea is so narrow that it can easily be blocked by a land force stationed at Sharm el Sheikh. In the case of an interior state, which is the third possible location, the co-operation of neighbouring states will be required, and if the aggressor is not a neighbour the policies will have to be operated by proxy by the neighbouring states; this is the situation in Rhodesia. When the country being blockaded is an enclave of the aggressor state the situation is much easier; Lesotho would provide the best example of this situation. As a general rule it seems likely that the problems of a successful blockade will increase with the number of countries which have to be involved; this is certainly the case in the sanctions against Rhodesia where South Africa and Portugal have declined to comply with the United Nations instructions.

Location, the nature of the coastline and accessibility to the hinterland, will also influence the extent to which alternative routes can be developed by the blockaded state. For example, Russia and Canada have long coastlines, but for part of the year these could only be used through the aid of a continuous programme of ice-breaking; Algeria has contact with five states in the Sahara, but the development of these as alternative routes would be both difficult and expensive.

The second major group of geographical factors will relate to the nature of the country's material and human resources. Enough has already been written to show that the ability of states to survive a blockade will vary with the resources they possess, including any stocks which may have been accumulated, the capacity of the state to increase the output of items which can no longer be imported, and the ability of the state to substitute alternatives for scarce items. It will also depend upon certain qualities of the population. For example, the strain of an economic blockade would probably create greater problems for a state with a heterogeneous population, where the differing sections had different attitudes towards the blockade, and perhaps felt the impact of the blockade in varying degrees, than in a state with a homogeneous population structure, which also had a complete identity of purpose and resolve. The preparedness of the people to accept lower living standards may also be a significant factor. For example, when Indonesia imposed a trade boycott on Malaysia during the period of confrontation there were predictable unfavourable reactions upon the Indonesian economy. The situation was prevented from becoming disastrous by the ability of the people to retreat into a form of subsistence economy. Such a development would be impossible in certain countries which depend on manufacturing industry as the main means of employment, and where there is a shortage of land. Hong Kong would perhaps be the classical case of this situation.

It is pointless to try to classify states according to their ability to resist blockades, or to construct models of all the theoretical possibilities, for each state will be original, and states would fit into different categories depending on which state was attempting the blockade, the attitude adopted by neighbouring states, and the general level of world trade, to name only three of the many possible variables. Jones (1954) has given an excellent account which provides a useful blueprint for geographers attempting such an analysis for a particular state in a particular situation. The important point to stress is that geographers have a major contribution to make in any assessments involving blockades, boycotts or embargoes, for they have a clear general understanding of patterns of trade and movement, resources and population structure which are important factors. It is also important to stress that geographers must concern themselves with identifying and measuring the changes which flow from the unilateral dislocation of trade and intercourse, distinguishing the temporary from the permanent, comprehending those which were expected by the planners and those which were unexpected.

Mobilisation of population resources

The aim, in policies of this nature, is to make the most efficient use of available population, so that the needs of the armed forces are met, both in terms of the number of servicemen, and in the employment of civilians to provide war materials and man the country's organisation. This has not always been the case, but the damages of a *laissez faire* policy to this problem are obvious. First, the voluntary enlistment of servicemen and women, without regard to essential occupations, may create labour shortages in vital industries. Parker (1957), writing about the situation in 1914, in Britain, noted this problem.

> But before the first winter of the war was half over the dangerous consequences of uncontrolled volunteering had become only too apparent. The munitions industries had lost many of their skilled and experienced workers and the output of equipment and weapons was already lagging behind the requirements of the Expeditionary Force and the newly formed contingents at home.
>
> (p. 1)

In the Second World War, it has already been noted that the surprisingly high rate of enlistment by farm labourers threatened the projected targets of production, and persuaded the government to introduce legislation to limit this drift. The second problem may be caused by a movement of population to safer areas of production, or to areas of labour shortage, where wages are higher, which could both create unused capacity in the areas of emigration. Parker (1957, pp. 123 and 127) noted both situations in Britain during the Second World War.

The problem of employing labour efficiently has many conditioning factors, some of which include questions of machinery, wage policy and labour relations, which are marginal to the interests of the political geographer. One of the more critical groups of factors, however, includes the distribution of labour and the distribution of industry. These are the initial patterns with which the government is faced, and which it must rearrange to suit its purpose. While labour is more mobile than plant, it is not completely mobile, and in any case the government can change the distribution of industry by moving existing machinery, and through control over the establishment of new industries. It is therefore basic that governments will normally exercise control over both plant and labour.

In respect of labour the major problem is to re-employ persons who were unemployed, or whose occupations, such as catering and house-building, are no longer available, and to control enlistment so that production in vital industries is not impaired. In many cases, labour needs cannot be satisfied from local sources; for example at the beginning of wars the tourist industry is immediately depressed, but areas where this labour becomes available rarely include centres of major industrial production. Or it may happen that the pools of unemployment exist in vulnerable areas where the government decides it would be unwise to concentrate further production.

Different states will adopt different combinations of policies to serve the satisfaction of labour needs, but in each case the first action will be to identify the areas of surplus labour and labour shortage. Such areas will be broadly predictable providing there is an understanding of where categories of industries exist and what numbers of workers are employed there; geographers would seem to be well equipped to play a useful role in the preparation of such inventories. In the Second World War the British Ministry of Labour classified areas according to the labour situation.

> 'Green' areas were those in which there was available labour beyond local needs that could be transferred to war production elsewhere. 'Amber' areas were those in which there was no marked surplus or deficiency of labour. Areas where there were urgent unsatisfied demands for labour were marked 'red', and areas where potential requirements were so great that no further production should be introduced if it could possibly be avoided, were painted 'scarlet'.
>
> (Parker, 1957, p. 148)

The plan was that industries producing necessary items for home consumption should be concentrated as far as possible in the green and amber areas, away from the main centres of munitions production in the red and scarlet zones. This meant that non-essential factories in the red and scarlet zones should close down and release labour for more vital employment. This policy for the concentration of industry was not a noticeable success. Firms which might have closed down were naturally reluctant to hand over their trade secrets to their peacetime competitors located in green or amber areas, and were fearful that they would not regain their identity after the war. Further, some industries consisted of a large number of small firms or a few large firms, in widely scattered areas; both situations raised problems of organisation. There was also a problem to find a sufficient number of suitable factories in the green and amber areas to maintain a satisfactory level of production for the domestic market. The most successful concentration occurred in those industries which were strongly organised and already localised. In industries such as pottery, cotton and hosiery and lace, the withdrawal and re-employment of labour was quite orderly. Parker (1957, p. 194) estimates that the total number of workers made available by concentration processes was 110,000–120,000.

Governments may also control labour to some extent during peacetime, when there is recognised a threat to state security. Mr Hasluck, the Australian Minister for External Affairs, writing in *Foreign Affairs*, noted that due to changed tactical circumstances, specifically the nature of modern conflict and the reduction in Britain's capabilities east of Suez, Australia could no longer rely on the traditional method of raising forces in time of war, but must maintain a larger standing army, supported by conscripts. In some Communist countries, such as Cuba, China and Russia, there is a much greater level of labour control at all times than in non-Communist countries. In Israel some type of national service for men and women has always formed an important part of national security.

Lastly, it can be noted that in cases where a state acquires authority over a hostile population during war, it may also seek to direct the production of that population. This point was well illustrated by German policies towards foreign labour in Europe, during the Second World War, which caused major changes in population distribution.

The government can influence the location of industry, first, by controlling the siting of new factories, and second by encouraging the selective growth of existing establishments through the placing of contracts. Hornby (1958, chapter IX) has noted that in Britain even before 1939, the service departments placed as many as possible of their contracts with firms in the depressed areas of Clydeside, Tyneside and Northern Ireland. He also describes the 1934 classification of Britain, by the Air Council, into 'relatively safe', 'unsafe' and 'dangerous' areas, based on their vulnerability to air attack, and shows how this classification had to be modified as the range of aircraft increased. While some industries such as aircraft construction were dispersed from Croydon, Southampton and Weybridge to South Wales, the upper Thames valley and Hampshire respectively, this was not the general rule. If industry was dispersed into rural areas it reduced efficiency, made defence more difficult, and rendered them more likely to detection by air photographs. Proximity to management was considered by the British Secretary of State to be the main factor determining new general locations. Another important factor concerns the nature and efficiency of transport routes within the country, and the extent to which old sources of supply can be maintained, or new ones have to be developed.

This study of the geographical factors which influence the formation of policies governing the mobilisation of material and human resources, and the identification of long- and short-term effects of the operations of such policies, provides a common interest for political and economic geographers. The political geographer is likely to find the statistical tools of the economic geographer most useful in measuring and comparing the changes which result from the operation of policies.

The acquisition of territory

To the present point, with minor exceptions, attention has been focussed on economic policies by which the state seeks to strengthen itself and weaken the enemy, both actual and potential. It is now necessary to consider policies of territorial expansion, which, whatever the particular motives, will usually serve both general aims simultaneously. Territorial expansion may be accomplished through conquest or cession. Conquest means the physical occupation of the territory of some other state by an act of war, or the deliberate movement of population into unoccupied or unclaimed land. Cession involves one state yielding a section of territory which it controls to another state, and this may be achieved by diplomatic, economic or military pressures which stop short of war, as well as by military success. But the distinction between these two methods is not a clear one, nor is it very important to political geographers; what is most important is the aim of one state to gain more territory at the expense of the other and the consequences which flow from this transfer of land from one political authority to another.

In examining this question geographers are concerned to identify the geographical motives of the expanding state, the geographical factors which may influence the timing of this programme, the geographical arguments which may be used by states as justification for such policies, the geographical factors which may influence the adoption of grand strategy and local tactics, the geographical basis of any agreements which may be concluded between states at the conclusion of such programmes of expansion, and the geographical consequences of such agreements. These interests have much in common with those outlined in the analysis of territorial disputes (Prescott, 1965).

The geographical motives of states seeking further territory may be varied. Some will seek to regain territory which they had lost, or to which they believe they have a right, or which they believe would be more appropriately included within their boundaries. China's reoccupation of Tibet, and Somalia's attempts to gain territory from Ethiopia and Kenya are apparently prompted by this motive of reconquest, while Morocco's claim to south-east Algeria also draws strength from the Moroccan belief that the Franco-Moroccan boundary lay east of the present line, and that the area can be more effectively developed as part of Morocco, than as part of Algeria. Some states will embark on a programme of expansion because they fear the dangers of the present situation. Peltier and Pearcy (1966, p. 41) may have had such a situation in mind when they wrote:

> . . . military war may be thought to result when one of the parties determines (1) that the national interest involved manifests the greatest national importance.

Israel has twice demonstrated such a situation in 1956 and 1967. The 1967 situation is clearest, because the 1956 situation was complicated by the nationalisation of the Suez Canal and charges of Anglo-French collusion with Israel. On 18 May 1967 the United Arab Republic requested the withdrawal of United Nations peace-keeping forces from Egyptian territory. This move was completed within four days. On 23 May 1967 Egyptian troops reoccupied the fort of Sharm el Sheikh, at the mouth of the Gulf of Aqaba, and it was announced that the gulf had been closed to all Israeli ships and all ships carrying strategic material to Israel. Soon after this Jordan and the United Arab Republic, whose governments seemed to have been hostile to each other, signed a defence treaty. On 5 June Israeli forces attacked, allegedly in response to an Egyptian attack, and in a few days they had captured the Sinai peninsula up to the east bank of the Suez Canal, the west bank of the Jordan river, and the strategic heights of Syria which commanded the Israel plains. At the beginning of the fighting Israel leaders announced that they had no territorial ambitions; within a short while all the leaders were agreed that Jerusalem ought to be retained, and some began to urge that not all the Arab territory should be returned, but that Israel should use the opportunity to fashion boundaries which would be more easily defended, and to secure access to the Red Sea.

The economic motive has also been the spur of territorial expansion. This motive was evident in certain aspects of Germany's campaign against Russia and Rumania. Medlicott (1959, pp. 641 et seq.) has suggested German concern with Swedish iron ore during the Norwegian campaign, with Rumanian oil in their Balkans campaign, and with Russian oil in their diversion of forces from Stalingrad to win the north Caucasian wells. There are many historical examples of states in Africa and Asia conducting wars which resulted in temporary territorial expansion, which aimed at the acquisition of slaves and food. In more recent times the most spectacular demonstration of expansion for economic gain has been found in the colonial policies of the major European powers. Economic gain was not the only motive, for strategy played a major role in a number of cases. The colonies sought for strategic purposes alone were usually small, and located at vital staging points or sites which would control narrow but important waterways. Colonies such as Aden, Gibraltar and the original Cape of Good Hope settlement are typical of such colonies.

Colonial policies have been of such importance in the past, and have such an importance in understanding the present political patterns of the world, that attempts have been made to raise it to the level of a separate branch of geography. In fact the study patently includes aspects of political, settlement, economic and historical geography. It is suggested here that the political geographer should focus his attention on the four major stages of colonial policy: acquisition, administration, disposal, and relations with the newly independent state. Pounds (1963, chapter 13) has written a very interesting account with a multitude of examples of the political geographer's interest in the first two stages, and these are generally the stages which have been most considered by political, economic and historical geographers.

As Pounds and others have pointed out, geographers' interest in policies of acquisition lies in discovering the geographical motives, in tracing the competition between rival powers for various areas, and in understanding the geographical factors which have influenced the colonial pattern. The major aspects of colonial administrative policy relate to the way in which geographical factors have affected the formation of such policies, factors such as the presence or absence of an indigenous society with a well-developed political system, or the suitability of the area for expatriate settlement, and the way in which the colonial policies have influenced the development of the colony, the development of relations between the colony and the imperial power, and the relations between imperial powers.

The present period is ideal for the study of the disposal of colonies, and for comparison with present experiences in disposal and past experiences, as in the last century in South America, or the early part of this century in the case of Turkey. Colonies may be disposed of in three ways: they may be incorporated into the imperial state, as in the case of Surinam and the Portuguese and Spanish overseas provinces; they may become independent states; or they may be transferred from one tutelage to another. States will dispose of colonies if they are unable to hold them in the face of military defeat, as in the case of Germany at the end of the First World War, or if the advantages of trying to hold them is outweighed by the disadvantages, or if the colony loses its particular value. The disadvantages of trying to hold Algeria by force were seen by General de Gaulle to be less than the advantages which the tenuous measure of control offered, and there was a rapid French withdrawal; Britain's revaluation of defence policy, and the new means of carrying out that policy east of Suez, deprived Aden and Cyprus of their previous importance as military bases and staging posts. Long-range, larger transport planes and amphibious task forces, which can make use of tiny island bases such as Gan and Aldabra, were significant in this revaluation.

Lastly there is the aftermath, when the colonial territory

becomes independent, or is absorbed, or is transferred. After independence the study of relations between the imperial power and its former colony can be studied in the same way as the relations between any two independent states, but their former connection will provide a vital factor in making the relations more or less cordial and in preserving or radically altering the previous patterns of contact and trade. The political geographer must also reassess the global patterns of political authority as the result of the formation of new states. If a state is absorbed, the follow-up study will examine the extent to which the new addition becomes integrated within the body of the state, and estimate the ways in which the union increases or diminishes state strength, and creates or solves political geographical problems. In the case of a colony being transferred political geographers will search, together with economic and settlement geographers, for the ways in which a change in political authority brings changes in landscape and economic development in train.

After this digression to present a unified concept of colonial studies in political geography, it is necessary to return to the way in which geographical factors influence the timing of plans for territorial expansion. It is axiomatic that a government will pursue policies of expansion when it believes itself to possess the greatest relative advantage. Claims to territory are frequently launched when the state on which the demands are made is in a weakened or divided condition. For example, the British advance from the foothills of Assam to the position of the McMahon Line occurred when the Chinese revolution had weakened Chinese authority in Tibet (Lamb, 1966). Rumania was subjected to claims by Hungary to Transylvania and by Bulgaria to the Black Sea Province of Dobrudja when Russia had advanced into its northern areas, and Germany was exerting pressure with regard to its oil supplies (Gwyer, 1964, pp. 57–9). Ward and Gooch (1923, vol. 2, pp. 342 et seq.) note that the Russian decision to attack Turkey was encouraged by the fortunate concurrence of rebel action in Montenegro, the dispute over the keys of the Holy Place, the demands of the Greek Orthodox Church for deliverance and minor problems in Turkish Asia. The timing can also be dependent upon simple geographical facts. British Intelligence officers had predicted that the attack by the Boers in 1899 would not be launched before October, because of the need for sufficient growth of fodder for their horses; the attack was launched on 10 October (Report of His Majesty's Commissioners, 1903,

p. 28). It is not contended that this was the only factor, but it certainly was an important factor. In other cases action has been postponed until the aggressive state had accumulated stocks to withstand any shortages which may develop after the onset of hostilities.

Arguments used to justify a war or claims to territory may have a variety of geographical bases as the following list shows.

(1) The need to acquire territory vital to state security. Israel undoubtedly regards certain areas conquered in June 1967 as being vital to her future security. In 1870 the North German Confederation announced that it would demand the cession of Alsace and part of Lorraine. The claim was justified in the following terms:

> It is not to be for the mere augmentation of German territory: nor is it for the purposes of improving the facilities for an attack by Germany on France. It is to be a defensive acquisition, and is simply to make it more difficult for France to attack Germany.
>
> (Temperley and Penson, 1928, p. 324)

(2) The need to unify areas occupied by fellow nationals. The statement by the Somali Government quoted above (page 70) illustrates such an argument.

(3) The need for access to areas which can absorb or feed a critical surplus population.

> The Japanese were concerned over the food supply at the beginning of the twentieth century. The shortage of agricultural products was considered the most alarming consequence of our population. The government feared that in time of war the people would be underfed and therefore Manchuria was regarded as being the most suitable source for nourishment of a prolific race. Here was found wheat, soya beans, peas, sugar, kaoliang, maize and millet.
>
> (Ennis, 1948, p. 304)

(4) The need for access to the sea or a navigable river. Pounds (1963, pp. 237–44) gives many examples of such claims and arguments.

(5) The right to occupy areas which were wrongly detached from the state at some time in the past.

> The wrong done to France by Germany in 1871 in the matter of Alsace-Lorraine, which has unsettled the peace of the world for nearly

fifty years, should be righted, in order that peace may once more be made secure in the interest of all.

(President Woodrow Wilson's Eighth Point)

The French refused to allow the return to be contingent on the vote of a plebiscite and they were right. In M. Pichon's words 'The question of Alsace-Lorraine is a Question of right'.

(Temperley, 1920, vol. 2, p. 167)

Geographical factors are often of prime importance in determining the overall strategy of a state's war aims, and are invariably very significant in determining tactical situations. Certain extra-territorial areas have always been of vital concern to individual states; this is demonstrated by Russian interest in the Baltic and Dardanelles, by American interest in South America and by British interest in the opposite coast of Europe.

It was the veriest commonplace of history that this fragment of the continent had for centuries been the prize for which endless wars were fought, and that the French were the chief aggressors in these wars, and that it was essential to the security of the British Isles to prevent any Great Power, most of all France, from holding the coast of Flanders, with Antwerp and the mouth of the Scheldt. Therefore, partly with the British object in view, but chiefly in order to erect a solid barrier on the eastern boundaries of France, the Allies, led by Great Britain, reunited the Southern and Northern Netherlands, which had been separated since the close of the 16th century. They [Castlereagh and other ministers of the Coalition] believed that the creation of this strong buffer would at least diminish the secular struggle for supremacy in the Low Countries.

(Ward and Gooch, 1923, vol. 2, p. 159)

Compare part of the speech by Sir Austen Chamberlain in 1925:

All our greatest wars have been fought to prevent one great military power dominating Europe, and at the same time dominating the coasts of the Channel and the ports of the Low Countries.

(House of Commons Debates, 5th Ser., vol. 182, p. 315)

Yet the fact that a government believes that it possesses a vital interest in a particular area doesn't mean that the policies towards that area will always be constant. Russia's policy towards Turkey in the nineteenth century was ambivalent. When Russian influence was paramount, the Czar resisted any tendencies which might lead to the break up of the Turkish Empire, but when Russian influence was ineffective, Russia at different times proposed the partition of the Empire to Britain, and sought to detach the Balkans to gain access to the Dardanelles (Ward and Gooch, 1923, vol. 2, p. 342). Technical changes may also allow changes in policy. The decision of the British Government to withdraw troops from its base at Singapore by the mid-1970's does not reflect a diminished interest in South-east Asia, but a conviction that the same influence for stability can be exerted by troops based either in Australia, Hong Kong, or in Britain, with rapid means of air transport and suitable staging points in the Indian Ocean. The need for economy in British defence spending may also play a role in this decision.

The importance of geographical factors in making tactical decisions is too well known to need elaboration. This is the main part of the field of military geography which has a long history, including works by Maguire (1899), May (1909), Macdonnell (1911), Johnson (1917), Clausewitz (1950) and Cole (1950). The latest contribution to this field is by Peltier and Pearcy (1966). This book has been given a mixed reception by reviewers, of whom Marshall-Cornwall (1967) and Jackman (1967) are representative of contrasting opinions; but perhaps the main significance of the book lies in the expanded claims to a particular field made by the authors. It is claimed that military geography is concerned with the 'interaction between geographic conditions and military affairs' (p. 18), starting with grand strategy and passing through military strategy to military tactics. Other subjects included are logistics, the influence of geography on military-civil relationships in occupied territory, and the division of the world into military regions. This expanded form is justified as the result of 'modern development of the concept of warfare and the expansion of areas of military interest and responsibility' (p. 166). For Peltier and Pearcy it is only in the area of 'national strategic considerations that military and diplomatic affairs merge into a relationship and that the distinction between military geography and political geography becomes difficult to make' (p. 8, see also p. 20). This would also presumably be true of geopolitics which is defined as 'the geography of political relations' (p. 136). This conclusion can only rest on a narrow definition of political geography and it is hoped that the foregoing discussion has shown that policy decisions about military matters are not alone in owing much to the influence of geographical factors, and that they form a continuum with other policy decisions to strengthen the state and weaken an enemy. In view of the concept of total war, Peltier and Pearcy should have given more consideration to economic warfare. Further, political geographers are concerned with the consequences of war, with the rearrangement of boundaries, the redistribution of population, and the development of new resources, to name only three aspects, giving a further contact with military geography and making the two indivisible. There is no reason why political geographers should not focus their attention on a single aspect of their subject, but they only do the subject a disservice if they attempt to construct these particular aspects into separate branches.

The significance of geography in the determination of any eventual peace agreement is the next facet of the study. This will be revealed through the provisions of the peace: the rearrangement of boundaries; the destruction of fortresses; the establishment of an occupying force; the payment of reparations in cash or kind or territory; guaranteed access to particular areas or particular routes. The rearrangement of boundaries after a war is a common occurrence as Europe has demonstrated twice this century. The new alignments will normally be designed to prevent a renewal of conflict by eliminating minorities, by removing dangerous salients, or by reducing the territorial resource base of an aggressor. They may also be made to provide compensation for war damage or to provide improved communications for one of the successful states (Prescott, 1965, pp. 120–1). Examples of peace treaties arranging for the destruction of fortresses are provided by the Treaty of Utrecht, which stipulated that the fortifications at Dunkirk should be razed and the harbour filled in, and the Younghusband Treaty between Tibet and British India in 1904, when Tibet undertook to dismantle forts on the road from India to Gyantse (Lamb, 1966, vol. 1, p. 244). It is interesting that in the former case the French produced a better harbour at Mordyke linked to Dunkirk by a canal (Ward and Gooch, 1923, vol. 1, p. 50). The right to station garrisons in Germany was conferred on the successful powers at the end of both wars in the present century. An older example is provided in a series of 'Barrier Treaties', 1709–15, by which the Dutch were given the right to garrison seven to nine strong places which included Namur, Tournai, Ypres and Venloo in the Spanish Netherlands (Ward & Gooch, 1923, vol. 1, p. 54). One part of the peace arrangements between Spain and Britain in 1713 stipulated that there would be no transfer or sale of Spanish land in America to France or any other country, providing an illustration of how a defeated state may have to accept limitations on policy for a period after the war.

Having considered the formation of policy—whether or not to attack or exert pressure to gain territory, when to attack, where to attack, when to make peace, the conditions of peace—it is necessary to examine the geographical results of such action. First it is necessary to measure the extent to which prediction matched performance, then there is the need to discover the unpredicted or unpredictable geographical consequences of the actions, which will include aspects of new political patterns at one end of the scale and landscapes altered in minute detail at the other end. The conduct of war is not related to the same economic principles which operate in peacetime, and in consequence change is often abrupt and in a different direction to the peacetime trends. War may represent a political or historical unconformity; marking dramatic change in a short period, with little or no apparent relationship between pre-war and post-war trends in volume of production, patterns of trade, and location of activities. Some influences will be short-term allowing earlier patterns to be quickly re-established. This was true when the Suez Canal was closed in 1956 during the Arab-Israel conflict. It will be interesting to see whether the Canal can regain its former importance after the 1967 closure, or whether there will be a major change in patterns of ship movements. The many papers dealing with geographical changes in war-affected regions, which appear after each major conflict, indicate that geographers appreciate the importance of such studies.

Alliances

To this point attention has been directed to the activities which a state can undertake alone if necessary—greater self-sufficiency, mobilisation, the use of alternative resources, blockades and invasion—but in many cases a state increases its strength through association with other states, and conversely weakens an enemy by alienating their allies. Most of the major textbooks in political science contain sections dealing with alliances because of the part they play in the balance of power between states (Morgenthau, 1960, chapter 12; Hartmann, 1957, chapters 16 and 17).

> The historically most important manifestation of the balance of power, however, is to be found not in the equilibrium of two isolated nations

but in the relations between one nation or alliance of nations and another alliance.

(Morgenthau, 1960, p. 181)

An examination of such texts shows that political scientists are concerned with alliances as a technique or strategy by which the state may defend itself or prepare the ground for an attack, they are concerned at the way in which the nature of alliances has changed in time, and the influence which alliances may exert on political developments in individual states. It follows that this interest in the subject of alliances is much wider than that of the political geographer. The geographer's interest in alliances relates first to the geographical factors and motives which encourage states to make alliances or to dissolve existing attachments, second to the area within which the terms of the alliance operates and their duration, and third to any geographical consequences of the treaty's existence or implementation.

It follows that an alliance must be based on a mutual interest and that the opportunities and gains for each partner must outweigh the responsibilities and disadvantages. There are several possible geographical motives in concluding an alliance. First, there may be the desire to augment both material and human resources so that any planned attack can succeed more quickly, or so that any threat may be defeated more certainly. This is surely the basic motive of many of the major alliances today—NATO, SEATO and the Warsaw Alliance. An example of aggressive intent is provided by the secret protocol attached to the non-aggression pact between Russia and Germany in 1939:

. . . the spheres of influence of Germany and the U.S.S.R. shall be bounded approximately by the line of rivers Narew, Vistula and San.

(Quoted in Hartmann, 1957, p. 98)

The Arab alliances against Israel implied the conquest and alteration of existing territorial arrangements. Alliances may also be concluded to secure one flank of a state while a conflict is waged on another. The German-Russian non-aggression pact served this purpose, and a similar aim was satisfied by the clause in the Triple Alliance of 1882, which stipulated that if either Italy, Austria-Hungary or Germany was attacked by a fourth power, the other two would maintain a benevolent neutrality at the least.

The danger which Germany temporarily avoided by its pact with Russia in 1939 had been the main intention of a Franco-Russian treaty in 1894.

> Article 1.
> If France is attacked by Germany, or by Italy supported by Germany, Russia shall employ all her available forces in order to attack Germany. If Russia is attacked by Germany, or by Austria supported by Germany, France shall employ all her available forces in order to fight Germany.
> (Quoted in Hartmann, 1957, p. 344)

Article III stipulated that France should supply 1·3 million men and Russia 800,000 and that these troops should 'engage fully and with all speed, so that Germany may have to fight at the same time in the East and the West' (quoted in Hartmann, 1957, p. 344).

Alliances may be forged between partners of disproportionate power to provide security for the weaker state, and to afford the stronger power the opportunity of operating its foreign policy at some distance from its shores, either through the right to intervene in any struggle, or the right to station troops at nominated bases. This has been an evident American policy in respect of Russia and China.

It is important to establish, if possible, the area to which the treaty applies and its time span, for these qualities will influence any geographical significance which the alliance may have. This may not always be possible, because treaties are often kept secret except in broad outline, and the general language may allow a number of interpretations. During the period of Indonesia's confrontation of Malaysia, there was a lively debate in Australia on whether the ANZUS Treaty committed America to support Australia if Indonesia began to infiltrate into Australian New Guinea. Hartmann (1957, pp. 290–1 and 313) has noted that the nature of alliances has changed with time. In previous periods they tended to be temporary, to meet a particular crisis in a particular area, and to end as soon as that crisis had passed. This was partly a consequence of the greater remoteness of areas, and the greater protection of physical barriers, which promoted a large measure of unilateral action and self-reliance by states. Today, with the greater efficiency of transport and weapons, the areas to which alliances apply have been greatly extended. But circumstances change and the new patterns of power may make old alliances redundant. The words of Lord Palmerston are probably still true for all states today.

> It is a narrow policy to say one country or another is a perpetual ally or enemy—our interests are eternal.
>
> (Ward and Gooch, 1923, vol. 2, p. 160)

Turning lastly to the consequences of alliances, it is clear that if the terms of the alliance are implemented and a war develops, then political geographers are concerned with the geographical consequences of that war in the way outlined earlier. However, the alliance may have certain geographical effects without the ultimate development of war. The effects may be in terms of the landscape—the establishment of bases by a stronger power, as in the case of the American-Thailand alliance, or the construction of roads and airfields. These specific and direct consequences may set in train a number of subsidiary effects relating to changed land use around bases, new patterns of civilian employment, and the resiting of settlements near new roads. It is also worth noting that the dissolution of alliances may produce certain geographical effects. When Britain announced that the Singapore base would be vacated by the mid-1970's, the Minister of Defence explained to a meeting of the backbenchers in his party that the run-down could not proceed more quickly because of economic and social considerations. This is presumed to refer to the considerable number of Singapore citizens employed at the base, who represent 10 per cent of the employed population (London *Times*, 20 July 1967).

If a basic strategic situation encourages two or more states to establish an alliance it is quite possible that closer economic ties will be encouraged.

Co-operation between states to secure alternative routes

Having examined the co-operation between states to augment their material and human resources, it remains to look at the special case of states which seek to decrease their dependence on a particular route, by developing alternatives through other states. It follows that such states are normally land-locked, and Africa, which contains the largest number of land-locked states provides examples in the past and the present.

In the second half of the nineteenth century the only land-locked states with a significant measure of autonomy in Africa were Ethiopia, the Boer Republics of the Orange Free State and the Transvaal. The Boer administration in the Transvaal was convinced of the need for an alternative outlet to that provided by the British Colonies of the Cape and Natal. There was a good cause for these views, for Robinson (1961) has recorded that British policy was based on three principles:

> . . . to exclude foreign powers from the coastline; to deny the Trekkers independent access to the sea; to ensure that the Colonies dominated the Republics.
>
> (p. 54)

The operation of this policy was revealed in the annexation of Natal in 1844 'to prevent the [Boer] Emigrants from ever acquiring a dangerous importance as an independent community . . . and to finally exclude the idea of foreign interference' (Uys, 1933, p. 10); the annexation of Zululand in 1885 to bar the Transvaal from Santa Lucia Bay; the annexation of Basutoland and Griqualand in 1868 and 1871 respectively; and the exclusion of the Transvaal from Bechuanaland by the London Convention of 1884. This latter area provides an interesting case of how quickly revaluation of territory becomes necessary as circumstances change. In 1882 Lord Derby, the Colonial Secretary, had announced in Parliament that:

> Bechuanaland is of no value to us . . . for any special purpose . . . it is of no consequence to us whether Boers or Native Chiefs are in possession.
>
> (Robinson, 1961, p. 203)

However, this view was overruled by February 1884, in the London Convention, when the Boers accepted that their western boundary lay east of the main route northwards through Bechuanaland. By the middle of 1884, when Germany had established a Protectorate over Angra Pequena, Bechuanaland had become a vital link in the British *cordon sanitaire* around the Boer Republics. However, the attempt to insulate the Boers from the sea was never completely successful, despite Rhodes' efforts in Rhodesia, for Portugal remained in control of Mozambique. Not that British interests refrained from planning to acquire southern Mozambique. There was one plan to secure it as compensation if Portugal could be persuaded to accept a loan and then default on the repayment, and during and after 1889 Rhodes tried to buy the whole of Mozambique south of the Zambezi (Robinson, 1961, p. 220).

The Boer plan to build a railway to Delgoa Bay was first the subject of a treaty with the Portuguese in 1875 (Van der Poel, 1933, p. 5), but problems of finance and administration prevented any progress. By March 1886 there was still no likelihood of the Portuguese link, so President Kruger proposed a customs union with the Cape and Natal. This was refused by the Cape Colony and four months later gold was discovered and Johannesburg became the new 'winning post'. President Kruger was now determined to keep out the southern rail connections until the link with Delgoa Bay was complete (Van der Poel, 1933, chapter 4).

> Every railway that approaches me I look upon as an enemy on whatever side it comes. I must have my Delgoa line first, then the other lines may come.
>
> (Robinson, 1961, P. 218)

In fact financial considerations prevented this plan from being successful; the railway from Cape Town reached Johannesburg in September 1892 (956 miles), followed in 1894 and 1895 by the Delgoa Bay (397 miles) and the Natal (494 miles) lines respectively. President Kruger attempted to ensure that the Portuguese line flourished by raising rates on the fifty-mile section from Johannesburg to Viljoensdrift on the State's southern border, from 2·4d per ton mile to 8d per ton mile (Van der Poel, 1933, p. 80), and by closing the drifts of Viljoensport and Zand on 1 October 1895.

Zambia's efforts to develop alternatives to the route via Rhodesia and Mozambique provides a contemporary example, although in this case the policy developed after many years of harmonious relations before Rhodesia declared itself independent. In fact the proposal for an alternative route was proposed before the Rhodesian rebellion seemed at all likely, but as this event began to seem probable, so the discussion became more detailed and the planning more urgent. In February 1964 the Zambian Minister of Transport announced that there would be no question of paying compensation to Rhodesian railways when the rail link with Tanzania, costing £17 million, was completed. A survey team provided by the World Bank reported two months later that the expense was not justified on economic grounds because of the low level of development in south-east Tanzania. In October 1964 discussions between the leaders of Zambia, Kenya, Uganda and Tanzania agreed in principle to build the link, which it was estimated would cost £50 million.

> The line would provide speedy transport of export commodities and encourage the fertile areas south of Iringa, the formation and expansion of dairy and horticultural industries for markets in Dar-es-Salaam and in Zambia. Steps are being taken to increase the output of the Zambia mines and it should not be difficult for Dr. Kaunda's Government to guarantee a high annual tonnage of freight for the line. The traffic alone might provide a reasonable measure of profitability while the additional revenue from the transport of agricultural products would enable the railway to provide a handsome addition to the overall income.
>
> (*The Standard*, Tanzania, 12 October 1964)

This seems to be a fair reflection of the views of the leaders at this time, although Uganda is reported to have had some doubts which were eventually overcome. In December the United Nations economic survey of Zambia warned that there was the danger of an expensive mistake over the construction of the link at that time. An inter-ministerial committee was established by the two countries concerned in May 1965, when Zambia's Transport Minister said, 'There is no doubt about the railway line. It is now a question of how and when the railway will be built.' In July 1965 the People's Republic of China offered £75–£150 million to build the railway, an offer which drew criticism from the Rhodesian and Portuguese Governments, whose countries would both lose revenue.

In addition to this major effort Zambia also canvassed the possibility of using on an increasing scale the railways which led via the Congo and Angola to Lobito Bay, and via the *voie nationale*, through the Congo to Matadi. The possibility of using the Benguela line was improved by the end of the Katangan secession, and the removal of the freight rates loading in favour of Mozambique ports. Until September 1965 the conference freight rates to Lobito had been much higher than to Beira and Lourenço Marques, even though the sailing distance from Europe is shorter and the delivery time to central Africa is thirty to thirty-five days less (London *Times*, 21 September 1965). The Benguela railway could carry 120,000 tons of copper from Zambia and provide a similar volume of imports, and this total could be greatly increased if the Cubal bottleneck could be eliminated; this occurs where the railway descends from the plateau over the coastal escarpment. Unfortunately for Zambia's plans there has been a persistent shortage of rolling stock, because of the Congolese Government's decision to send all its exports by internal routes to save foreign exchange. Efforts to provide an alternative road route have been much more successful. In May 1966 the Zambian and Tanzanian Governments announced the formation of the Zambian-Tanzanian Road Services Company to start operations in July of the same year. The Company will eventually have 430 trucks and trailers, each capable of carrying thirty-two tons. The schedule of charges relating to six categories of commodities varies from £15 per short ton of copper to £25 per ton for goods needing special treatment such as fruit and chemists' supplies (*Times of Zambia*, 7 May 1966). At £15 per short ton the road rate is only 10s per short ton more than the basic rate for transport by the Rhodesian Railways. However, the picture is less favourable when it is realised that the last 20 per cent of Zambia's copper export via Rhodesia is carried at the rate of £9 10s 6d. This discount was introduced in 1960 by the Federal Government to recapture the trade which was being legitimately carried by the Benguela railway. This means that if all the copper is exported via Rhodesia the average rate is £13 10s per short ton, a saving of 30s over road transport.

At the end of the first year of operation it was estimated that 25,000 tons of copper had passed through Dar-es-Salaam. By then 200 lorries were operating. Copper was also being carried by plane, which at first thought seems very uneconomic. The Roan Selection Trust Mining Group started using two American Hercules and it was estimated at first that they would carry 40,000 tons per year by means of a shuttle service. The actual rate has been below this because the planes have carried twenty tons per flight instead of twenty-five. The latest figures indicated an export by air of 1,800 tons per month. The Zambian Government announced in March 1966 that two BAC III jet transports would be purchased 'as one of the moves to make Zambia more self-sufficient'. Even if road and air transport operated at the highest possible level throughout the year, it is unlikely that the total volume of copper moved could exceed 72,000 tons; this is less than 11 per cent of Zambia's 1966 production.

This underlines the problem of developing major transport systems at short notice. Although the rail report, prepared by

Canadian and British officials, was presented to the Zambian Government in August 1966 no start had been made a year later. The report is believed to have indicated that there were no insupperable problems, but that the railway would have to cross a long section of slippery and treacherous soil. It is now (August 1968) more than four years since the idea was seriously canvassed and nearly three years since Rhodesia seized independence. Reports on the construction of the road from Iringa in Tanzania to Tunduma on the Zambian border indicate that it will take at least four years to complete. Until the road is tarred, surface conditions will reduce the volume of traffic which can be handled during the rainy season.

Clearly it is a sound policy on the part of landlocked states to develop alternative routes which can be used in times of difficult relations with one of its neighbours. Such alternative routes will be expensive unless they are used, and it is always possible that the construction of a new route will lead to retaliation through increased rates by the country controlling the old route. Each situation is unique, depending on the location of the country, its neighbours and the characteristics of its trade. High value, low bulk items, such as diamonds, can be cheaply exported by air, which is uneconomic for bulky items, except in times of unusually high prices. The decline in copper prices in 1966–7, from a peak at £469 5s per ton to £345 10s per ton, must have made Zambia's efforts to export via Dar-es-Salaam even more uneconomic. It will be necessary to study changes in transport techniques which influence the cost of movement and the ease of construction. For example, patterns of air traffic will change with the introduction of increasingly large transport planes. Already it is proving possible to move coal along pipelines, by crushing it and flushing it through with water; this technique may prove capable of further adaptation and wider use.

REFERENCES

AFRICA RESEARCH BULLETIN, 1965, *Economic, Financial and Technical Series*, Exeter.

ARFA, H., 1966, *The Kurds: an historical and political study*, London,

CLAUSEWITZ, K. von, 1950, *On war* (translated by O. J. M. Jolles), Washington.

COLE, D. H., 1953, *Imperial military geography*, 11th ed., London.

DERRY, T. K., 1952, *The campaign in Norway*, H.M.S.O., London.

EDMONDS, G. J., 1967, 'The Kurdish war in Iraq: a plan for peace', *Roy. Centr. Asian J.*, **54**, pp. 10–22.

ENNIS, T. E., 1948, *Eastern Asia*, New York.

GWYER, J. M. A., 1964, *Grand strategy*, vol. 3, pt. 1, H.M.S.O., London.

HAMMOND, R. J., 1951, *Food*, vol. 1., H.M.S.O., London.

HARTMANN, F. H., 1957, *The relations of nations*, New York.

HASLUCK, P., 1964, 'Australia and Southeast Asia', *Foreign Affairs*, **43**, pp. 51–63.

HASSAN, A., 1966, *The Kurds: an historical and political study*, London.

HORNBY, W., 1958, *Factories and plant*, H.M.S.O., London.

HURSTFIELD, J., 1953, *The control of raw materials*, H.M.S.O., London.

JACKMAN, A. H., 1967, review of Peltier and Pearcy (1966), *Professional Geographer*, **19**, pp. 50–1.

JOHNSON, D. W., 1917, *Topography and strategy in war*, New York.

JONES, S. B., 1954, 'The power inventory and national strategy', *World Politics*, **6**, pp. 421–52.

KEESING'S CONTEMPORARY ARCHIVES, 1967, Bristol.

LADAS, S. P., 1932, *The exchange of minorities Bulgaria, Greece and Turkey*, New York.

LAMB, A., 1966, *The McMahon line*, 2 vols., London.

MACDONNELL, A. C., 1911, *Outlines of military geography*, 2 vols. and an atlas, London.

MAGUIRE, T. M., 1899, *Outlines of military geography*, London.

MARSHALL-CORNWALL, Sir J., 1967, review of Peltier and Pearcy (1966), *Geogr. J.*, 133, p. 113.

MAY, Sir E. S., 1909, *An introduction to military geography*, London.

MEDLICOTT, W. N., 1959, *The economic blockade*, vol. II, H.M.S.O., London.

MINISTRY OF AGRICULTURE, 1926, *Agricultural policy*, Cmd. 2581.

MORGENTHAU, H. J., 1960, *Politics among nations*, New York.

MURRAY, K. A. H., 1955, *Agriculture*, H.M.S.O., London.

PALLIS, A. A., 1925, 'Racial migrations in the Balkans during the years 1912–24', *Geogr. J.*, **66**, pp. 315–31.

PARKER, H. M. D., 1957, *Manpower: a study of wartime policy and administration*, H.M.S.O., London.

PELTIER, L. C., and Pearcy, G. E., 1966, *Military geography*, Princeton.

POEL, J. van der, 1933, *Railways and customs policies in South Africa*, London.

POUNDS, N. J. G., 1963, *Political geography*, New York.

REPORT OF HIS MAJESTY'S COMMISSIONERS APPOINTED TO ENQUIRE INTO THE MILITARY PREPARATIONS AND OTHER MATTERS CONNECTED WITH THE WAR IN SOUTH AFRICA, 1903, Cd. 1789, H.M.S.O., London.

ROBINSON, R., Gallagher, J., and Denny, A., 1961, *Africa and the Victorians: the official mind of imperialism*, London.

ROSE, A. J., 1966, *Dilemmas down under*, New Jersey.

SEGAL, R. (Ed.), 1964, *Sanctions against South Africa*, London.

STATISTICAL DIGEST OF THE WAR, 1951, H.M.S.O., London.

TEMPERLEY, H. W. V., 1920, *A history of the peace conference*, vol. II., London.

TEMPERLEY, H. W. V., and Penson, L. M., 1938, *Foundations of British Foreign policy from Pitt* (1792) *to Salisbury* (1902), Cambridge.

UYS, C. J., 1933, *In the era of Shepstone*, Lovedale.

WARD, Sir A. W., and Gooch, G. P., 1923, *Cambridge history of British foreign policy*, 3 vols. Cambridge.

WHITTINGTON, G., 1967. 'Some effects on Zambia of Rhodesian Independence', *Tidjschrift voor Economische en Sociale Geographie*, **58**, pp. 103–6.

4

ADMINISTRATIVE POLICIES

This chapter is concerned with those aspects of the organisation of authority throughout the state which may be influenced during their creation or operation by geographical factors, or which may influence the geography of the state through their operation. Clearly this will exclude a large volume of legislation which is remote from geography such as the administration of justice. In a very narrow sense most administrative policies could be classified under the heading of policies of development, since they are presumably designed for the promotion of the state's welfare, and some of the remainder could be classified as policies of defence, especially those relating to important minorities as in Cyprus; however, it seems more useful to consider these policies together, because organisation in some form is a common feature of all states, and because similar arrangements in different states may be designed for different purposes. For example, the control over land which is exercised in the United Arab Republic and Rhodesia have different aims; in Egypt control is largely an economic device to ensure the best possible development according to the interpretations of the government; in Rhodesia the land apportionment provisions undoubtedly carry defensive overtones for the Rhodesians of European descent.

The administrative system of any country is usually contained in the constitution and specific laws giving effect to its provisions. In some cases, as in India, the constitution is a detailed document, designed to cater for all conceivable situations, in other cases, as in Israel, it is a short document, which lays down fundamental principles. But whatever its form, political geographers must be prepared to study the provisions of the constitution and the law which are relevant to the political geography of the state. Since the form of the administrative system and the way in which it is defined varies so widely throughout the world there is no short cut to the identification of relevant sections, it is simply a question of thorough reading. This suggestion is not novel for Whittlesey (1944) gave the clue many years ago.

> The political structure erected by every group of people is, ideally, a device for facilitating the economic and social life of the community. It is most successful when it neatly fits the conditions of the natural environment in the area where it functions.
>
> (p. 556)

> Every political system is the summation of laws which people make in order to extract a livelihood from their habitat. It follows that the political concepts held by any group are affected by their natural environment.
>
> (p. 557)

> Legal systems are images of the regions in which they function, sometimes faithful and sometimes distorted. Individual laws mirror the society and the habitat by and in which they are created. Because humanity occupies its habitat dynamically, laws tend to become outmoded. When this occurs they are occasionally revoked, sometimes they are disregarded, usually they are given a new meaning.
>
> (p. 565)

> The interplay of law and regions is incessant. Each affects the other and in the process is itself modified.
>
> (p. 589)

Whittlesey (1944) noted that there were few specific studies of the interplay of geography and the law when he wrote his book; this lack still exists today. Where he used the word 'laws' we can equally substitute 'constitution' or 'administrative structure'. Buckholts (1966) who has acknowledged his debt to Whittlesey makes occasional reference to the constitutions of states in his systematic study of political geography.

Although there is no short cut in the geographical analysis of constitutions and administrative systems, it seems probable that the relevant sections will be related to one or more of four topics: control of land, control of people, the administrative structure, and elections.

Control of land

The policies being considered here are those which lay down the general principles of land and resource control; the detailed policies which influence the use made of land will be considered in the chapter dealing with policies of development. While most constitutions contain some reference to the area of the state and the fact that none of it may be detached, some constitutions go much further and claim control for the state over all land and resources which may be derived from it. The Egyptian constitution claims for the state, control over all the natural wealth of the country, including 'the subsoil and the territorial waters'. Constitutions sometimes include references concerning the alienation of land. Anyone studying the political geography of African states must take into account the regulations governing the right of expatriates to purchase land. In Nigeria and Lesotho alienation was forbidden, in the latter case all land is invested in the king for the benefit of the Sotho people. In Kenya, Swaziland and Botswana alienation was permitted at various times, and as a result these countries have had to face problems which Nigeria and Lesotho were spared. The Kenyan situation has been well documented in papers by Morgan (1963) and Jones (1965). In Mexico the constitution lays down that no alien may acquire land within a zone stretching 100 km from the land boundaries and 50 km from the coast; it would be interesting to know just why these particular limits were fixed. In Papua and New Guinea, which are administered as a single area, there were different policies towards alienation of land during the period when Germany controlled New Guinea. As a result, while there is uniform control today through the Administrator of the territory, the freehold land which has been alienated in both cases varies. In Papua only 24,280 acres out of a total 55·1 million acres have been alienated on a freehold basis; in New Guinea 536,711 acres of the total of 59 million acres had been alienated on a freehold basis. There is a closer correspondence between the leasehold areas which are under expatriate control: 371,864 acres in Papua and 393,604 acres in New Guinea.

The third aspect of land control which interests political geographers occurs in those cases where the constitution makes provision for the allocation of land amongst various ethnic or racial groups within the state. The best examples of such cases today are found in southern Africa in South Africa, South West Africa and Rhodesia. In South Africa 264 areas of land totalling

57,933 square miles have been designated for occupation by Bantu only. These territories have been organised into a system of tribal, regional and territorial authorities which are eventually scheduled to receive self-government in respect of internal affairs. This system has now been adapted for South West Africa; the recognised ethnic groups, which have traditionally been located in distinct areas, will gradually be granted internal self-government until a point is reached when the groups can negotiate with each other over possible measures of political and economic co-operation. Most progress has been made in respect of the European, Coloured and Ovambo groups.

In Rhodesia the section of the constitution dealing with land is long and detailed (Federation of Rhodesia and Nyasaland, 1961; Rhodesian Government, 1965); there is no significant difference between the 1961 and 1965 constitutions. Tribal trust land, which is held on a communal basis by some Africans, is vested in a Board of Trustees which may vary the area controlled in accordance with certain provisions. For example, land can be withdrawn from this classification if it is required for defence or mineral exploitation, or if the tribesmen occupying an area wish to convert their rights to freehold ownership, or if it is decided that an area may be irrigated to the benefit of the tribesmen. There are conditions governing such withdrawals. In the case of mining and defence requirements the dispossessed tribesman must receive alternative land, suitable in quality to his agricultural or pastoral pursuits, or financial compensation; in the case of withdrawal for irrigation the land must be reserved exclusively for tribesmen.

The effects of these land policies can be readily discovered; they have an influence on the distribution of population, the development of communications, the establishment of industries and the movement of migrant labour. Brookfield (1957) has explored some of these influences in respect of South African and Rhodesian policies. It is more difficult to discover the actual geographical factors which influenced the adoption of these policies, although certain assumptions can be made. The enactment of these provisions is designed to prevent the territorial mixing of different racial groups from becoming more complicated, thus making racial policies more difficult to pursue. This is also demonstrated by the simplification of the racial distribution which has been undertaken in the major urban areas. To some extent the policy must be also designed to preserve a minimum area for the economically weaker group, which otherwise might be persuaded to dispose of land for money. Lastly the policy partly reflects the different systems of land tenure and attitudes towards land which are held by different racial groups. Hailey (1957, pp. 702–3) has shown that the present system in Rhodesia is descended from an original policy aimed at the creation of two consolidated African Reserves. In 1923, by which time Africans had purchased 100,000 acres in the European farming area, which was then a legitimate process, European farmers began to press for the complete separation of European and African land. This was recommended by a Land Commission in 1926, on the grounds that until the Africans were more advanced the points of contact between the two races should be reduced.

Control over people

Some of the policies in respect of land also exercise control over people by determining where they may live. Policies enshrined in the constitution and laws, which will be relevant at this point, are most likely to be found in those cases where the population is heterogeneous in certain characteristics, and efforts are either made to cater for these differences or eliminate them. They may be preserved through a system of applying different laws to different areas where the population groups occupy discrete areas, or to different groups wherever they may live, when the population is inextricably intermingled.

In either case, whether differences are preserved or eliminated, the policies are likely to be in respect of religion, language, legal system and representation. There are a number of states where there is more than one official language, without including the language of the former colonial power. The different areas of French and Walloon speakers in Belgium provides the best-known case, but problems of enforcing a uniform language in India and Ceylon have led to concessions to the Sikhs and Tamils respectively. In an earlier chapter it was noted that one of the successful demands of the Kurds in Irak was for the use of Kurdish in official business.

Different legal systems are applied in different parts of a number of countries. In the Middle East, countries such as Lebanon, Israel and Jordan have special religious courts which deal with matters of marriage, property and inheritance for different religious groups; in a number of African states, customary tribal courts still exist and administer systems of local justice. Variations in the system of representation to give recognition of minority groups are considered in the section on elections.

Administrative structure

There are two levels at which this particular enquiry should be undertaken. First there is the form of national government which may be federal or unitary; second there is the form of local administration.

Robinson (1961) has described a federal structure as 'the most geographically expressive of all political systems', and published work by geographers shows a keen awareness of this fact. In addition to specific studies of Australia (Robinson, 1961), Malaysia (Fisher, 1963), and Nigeria and Kenya (Prescott, 1958 and 1962), Pounds (1963) includes a survey of federal systems throughout the world. According to Wheare (1951, pp. 35–6) there are three prerequisites for a successful federation. First, the states should desire unity with each other; second, they should desire to retain some measure of regional autonomy; third, they should have the capacity to operate the federal system.

Geographers searching for geographical factors which were important in the decision to federate will usually find a multitude of reasons relating to defence, trade, economic development, access to the sea, complementary resources and a common ethnic origin. The geographic factors which may serve to persuade states that they need a significant measure of local autonomy will include remoteness, ethnic differences, and specialised economies which together with other factors may have fostered a distinct regional identity. The capacity to operate the federal system will be partially conditioned by the balance between the centripetal factors encouraging federal union, and the centrifugal factors promoting a desire for some regional autonomy. If either become disproportionately powerful the state will tend towards a unitary form or redivision into a number of separate states. The success of the federation may also be contingent upon the approximate equality in wealth, size and population of the various component parts. If there is a serious disproportion the chances of friction between the largest and smallest states will be increased. Nicholson (1954) has shown that boundary construction in western Canada was partly designed to make the units of equivalent size.

In recent times there has been a tendency on the part of British Governments to solve some of their problems of decolonisation through the construction of federations. The problems have been of two kinds. In states such as Malaysia, the West Indian Federation and Aden, the British Government sought to solve the problems of the small states which represent the flotsam and debris of empire. Sarawak and Sabah were not considered to be viable units and this, together with defence considerations, encouraged their final association with Malaya, and originally Singapore. In Aden and the British territories of South Arabia it was unthinkable to return the area to its condition of small Arab states and so a federation was mooted. In states such as Uganda and Kenya the initial quasi-federal structure was considered to offer the best chance of protection to tribal minorities. In both sets of cases, where the federal structure has resulted largely from an effort to solve a colonial problem, rather than a genuine desire firmly based in the geography and the politics of the countries for them to unite in this way, the experiment has not been markedly successful. The West Indian and Aden federations collapsed; Singapore withdrew from the Malaysian Federation; and the constitutions of Uganda and Kenya have been modified to a unitary form. Unfortunately, even when the political form is in harmony with the geographic and political environment, as in Nigeria, there is no guarantee that the federation will succeed.

There are three policy aspects of federations which will appeal to political geographers. First there are the geographical reasons which led the states to seek this particular political structure. Second it is important to look at the actual federal form selected. Pounds (1963) has shown that there is a considerable variation, varying from the almost unitary condition to the status where the individual components have very wide powers of internal administration. Lastly it is important to study the way in which the federal form changes, either in respect of the division of powers between the central and regional governments, or in the creation of new states or the seccession of existing states.

Turning now to the policy study of local government, it is useful to recall that Pounds (1963, p. 193) has rightly stressed that geographers must approach the study of politically organised areas from two directions: the division of territory and the division of power. These are the two means, usually related, by which governments arrange the control of their territory. Although Pounds devoted most of his attention to the territorial division, he acknowledged that the division of responsibility was also important. In future it might be worth while for geographers to pay more attention to the division of responsibility, for it follows that the political geographical significance of local government areas is not due only to the way in which boundaries are arranged, but also to the authority which *may* be exercised within those boundaries, and the way in which it *is* exercised. It is perfectly possible for significant changes to occur in the administrative structure, which are important to the geography of the state, without the boundaries being altered. Hall (1966) noted this situation in his study of local government in Japan.

> The most striking feature of local administration in Bizen of 1200 is that, despite successive changes in the organisation of power, the boundaries within which local administration was exercised tended to conform to the familiar shapes which had existed from the Nara period.
>
> (p. 158)

Political geographers are interested in the administrative structure of states for three reasons. First, as mentioned earlier, it is only through understanding the scope of authority within specific areas, and knowing the location of those areas, that the significance of such distributions can be measured, whether the study is concerned with comparisons of different levels of government within one country, or similar levels in different countries, or with the landscape effects of such systems. Alderfer (1964), a political scientist, has underlined the importance of studying local government systems in a way which geographers will understand.

> The form and structure of local government throughout the world today are the products of a history that covers many centuries . . . while there are many variations, there are no accidents; each and every beam or column has a meaning of its own . . . they tell about how things once were, or about what people thought should be, or perhaps what might be in the future if all went well.
>
> (p. 17)

Second, political geographers are interested in territorial patterns of power, because they are part of the contemporary political landscape which they are *obliged* to describe. Anyone who has tried to reconstruct the pattern of local divisions for a time long past, or a recent period when documentation is incomplete, will agree with this view. No political geography of the state can be complete unless it includes a clear statement of how the state's internal administration is organised. Fawcett (1961, pp. 33 et seq.) convincingly demonstrates the complexity of local government divisions in Britain and the importance of understanding them.

Third, political geographers are interested in this subject because it is one to which they can make a contribution when changes are considered. East and Wooldridge, in their introduction to their revision of Fawcett's *Provinces of England*, noted that he was convinced that if changes had to be made in the administrative structure, then 'the way to achieve them was to create provinces which were meaningful geographically'. The contribution which geographers can make to the definition of boundaries and the identification of regions has been considered elsewhere (Prescott, 1967, ch. 6); it is clearly much more difficult for geographers to act as experts in the question of dividing responsibility between government departments or between various levels of government. This difficulty has not prevented some geographers from making useful recommendations on the creation of bodies to supervise some aspect of resources control or economic development.

Finally it is worth remembering that studies of local government by political scientists frequently neglect the territorial aspect of the subject; the useful studies by Alderfer (1964) and Hinden (1950) are examples of this situation. This attitude towards the neglect is shared by McColl and Helin:

> Many scholars have tended to overlook the importance and significance of internal administrative changes in the development of nations.
> (McColl, 1963, p. 53)
>
> The mosaics produced by such administrative fragmentation have failed to capture the imagination and interest of American geographers.
> (Helin, 1967, p. 481)

It is now necessary to outline those aspects of the study which should be considered. First, it is important to establish as far as possible the purpose of any decision regarding alterations in the administrative structure. Fawcett (1961) believed that an essential assumption must be that any alterations are designed to facilitate good government. Lipman (1949, p. 1; see also Royal Commission, 1960, p. 59) spells out the twin basic aims of any system of local government in greater detail, which he maintains are the administration of public services for the benefit of all citizens, and the civic self-education of the population. It is pointed out that the first aim is best served by an arrangement giving maximum efficiency and economy, while the second aim demands that the administrative area should approximate to the social community. These criteria will not always be compatible. It must be noted that usually the only aim of special purpose regions will be efficiency and economy. But in addition to these basic aims, which should underlie every decision which alters an existing administrative structure or creates a new one, there may be specific aims. For example, Pounds (1963, p. 203) noted that in France arbitrary divisions were partly designed to discourage the further development of local patriotism. Hall (1966, p. 30) identified a similar aim in the regulations of the Ojin, at the end of the fifth century; he wished to limit 'the strong regional influence of the Kibi Chiefs'. While the preamble to bills altering any system of local government will probably set out the aims in general terms of 'good government', instruments creating special purpose areas will frequently be more precise about the specific aim.

Having identified the purpose of the change the political geographer's next concern will be to assess the role of geography in establishing the new system. Geographical factors may enter into the process in three ways. First, changed geographical patterns may make a review of an existing system necessary.

> The existing division of the country . . . dates back to a time when modern means of communication and transportation were lacking in Japan. It is said that the size of the prefectures were determined with the idea in mind that it should be possible for a man on horseback, setting out from any part of the prefecture, to reach the prefectural capital within one day. Clearly, this standard has little relevance today. The present prefectures also do not coincide with any areas that share common economic problems or needs.
>
> (Steiner, 1965, p. 151)

> While the actual distribution of these towns (in east England) dates back to their origin some six hundred or more years ago, the development of transport and of marketing organisation, and the growth of new social services, have occasioned the concentration of functions in a few, and this involves considerable extension of the areas which the latter serve. This maladjustment of present economic, social and administrative services to the distribution of towns, which is in the

main a medieval legacy, is reflected in . . . problems of rural activities and organisation.

(Dickinson, 1960, p. 89)

Similar points have been made by Fawcett (1961), Gilbert (1939), Whittlesey (1944) and Taylor (1942), suggesting that the geographical changes which will eventually make any system of administrative division obsolete will be mainly related to population movement, such as the growth of urban areas and rural depopulation, the development and exploitation of new resources, and the extension of conurbations beyond their legislative boundaries. This last development in the case of London was mainly responsible for the Royal Commission into this question which was created in 1957. Whittlesey (1944, p. 565) and Pounds (1963, p. 213) have noted that there is invariably a lag between the reason for a change and its accomplishment.

Geography may also enter into the process by being taken into account during the formulation of principles for areal division, and during the detailed application of these principles to the landscape. For example, Lipman (1949) has described the geographical principles which determined the definition of poor law areas in an abortive bill of 1880.

(a) No poor law parish or union shall extend over the boundary of any county.
(b) No poor law parish be divided into isolated parts.
(c) No poor law parish shall be too small or too great or having parts thereof so situate as to render the administration of relief to the poor in, or the local government of, such parish . . . in the opinion of the Commissioners inconvenient.
(d) Every highway parish shall be coincident in area with a poor law parish.
(e) Every highway district shall be coincident with some rural sanitary district.

(p. 135)

These principles in no way determine the actual alignment of the boundaries of these areas. During evidence given to the Royal Sanitary Commission in 1870 (quoted by Lipman, 1949, pp. 50–2), an inspector made it clear that the way in which the areas were determined was by the selection of a fitting centre, which considered the convenience of the governed and governors as far as possible. Steiner (1965, p. 24) has shown how after two Japanese rebellions in 1874 and 1877 a new system of local government was established on the principles that each area should have an output of 100,000 koku of rice, and that each prefectural capital should be easily accessible from all parts of the prefecture. Many government reports dealing with administrative reform outline the principles which have guided the changes, and the influence of geography can be detected. For example, in an appendix to the command paper on Local Government (1956) representatives of the local authorities associations outlined the guiding principles of boundary review. They are community of interest, development or anticipated development, economic and industrial characteristics, physical features, financial resources, means of communication and access to centres of administration, business and social life, area, shape, population size and distribution, the record of the administration and the wishes of the inhabitants.

Lipman (1949) includes a detailed note in his book on the way in which the division of France into *départements* was accomplished. The Commission of 1789 was apparently influenced by the egalitarian spirit of the day which demanded equality in size, the views of d'Argenson that the divisions should be as large as possible to secure efficiency without endangering the authority of the state, the scheme of Condorcet that each division should not exceed a maximum radius related to ease of transport and community of culture, customs and habit, and a map prepared by Robert de Hesseln, which for cartographical purposes was divided into eighty-one divisions measuring eighteen leagues square. The report of Thouret to the Assembly explained how the final decision to create eighty *départements* was implemented. The basic geometric pattern was distorted to take account of the cultural and physical features of the landscape, including the existence of former historical boundaries.

While many writers have stressed that it is the core of administrative regions rather than the boundaries which are important, clear boundaries are a great advantage in reducing friction and confusion, and the detailed examination of many local government boundaries in a variety of countries will show how they have been made to coincide with cultural and physical features, such as streets, railway lines and streams. Geographers such as Fawcett (1961), Gilbert (1948), Holmes (1944) and Taylor (1942) have offered sound advice on the construction of local government boundaries, which has been endorsed by Lipman (1949) and various authorities interested in this subject.

The next aspect of interest to political geographers is the actual pattern of administration which is established, which includes not only the pattern of areal division, but also the division of power at different levels of government. This is basic information about states which is a vital part of their political geography, and which is indispensable in any reconstruction of past periods. One particular feature of this aspect is the degree of coincidence between various categories of local authorities. Ullman (1939), Fawcett (1961), Gilbert (1939), Lipman (1949) and Holmes (1944) have illustrated examples where there was sometimes little coincidence between various boundaries, resulting in very complex patterns of territorial division of authority. Maps attached to the Report of the Royal Commission on local government in Greater London (1960, maps 4 and 12) showed the overlapping of local government and service boundaries in the capital. This is an aspect which geographers are well qualified to consider; they should not neglect the 'muddle of functions' resulting from overlapping authority which may be a significant factor in landscape development. Steiner (1965) has devoted a very interesting chapter to this problem in contemporary Japan.

This description of the patterns of power division leads to an analysis of how the existence and operation of these divisions influences the development of the landscape. This is a field of political geography which has been little explored judged by the number of published contributions; those by Nelson (1952) and Pounds (1963, pp. 203–6) are probably the best known. It should receive more attention in the future, not only because it is part of the total of political geography, but also because it is a valuable point of contact with geographers interested in other branches of the subject, such as settlement forms, land-use and historical development. The influence of different local authorities will probably be most apparent in urban areas, but the scope of investigation should be extended to rural areas, where different policies towards rateable values and drainage may create small though distinguishable differences.

It is also worth while to try to measure the extent to which the local population within any local area feels a sense of community or attachment. Pounds (1963, pp. 199–200) has recorded the contrast between the sense of loyalty in the United States, which is small at all levels below the State, with the situation in England, where parishes have been important, and where counties are still important today. There was certainly no shortage of local sentiment expressed during the recent reorganisation of boundaries in England (Prescott, 1967, pp. 173–4).

The last aspect of local government systems which concern political geographers is the ease with which they can be changed, which usually relates to the way in which they have developed. Steiner (1965, p. 35) draws the basic distinction between the system which has evolved from local sources and the system which is imposed by the superior authority of the state. It is likely that the former system will be more difficult to change and therefore liable to survive for longer periods, and will reflect more faithfully the patterns of local and regional sentiment throughout the state. While patterns which are superimposed by the state may show a great awareness of the geography of the state and the various needs of distinct regions, they are likely to be more easily changed, as the government discovers faults or disadvantages in the existing system. One of the recurring themes in the study by Hall (1966, pp. 29, 79, 158, 220, and 419) concerns the early establishment and persistence of certain areas in the Kibi area of Japan. This contrasts with the more rapid change in internal organisation within Russia and more recently the Congo (Kinshasa).

An earlier study (Prescott, 1967, pp. 178–9) suggested the need for a classification of internal boundaries; this seems less valuable than a classification of local government *areas*. It is doubtful, however, if elaborate and comprehensive classifications would repay the effort involved in their construction, due to the great complexity of local circumstances in the various states of the world. The most useful classification seems to be that between local government and special purpose administrative divisions described by Pounds (1963, chap. 8), which agrees with the distinction suggested by Fesler (1949) between governmental areas, possessing a measure of functional or fiscal autonomy, and field service areas designed for the convenient execution of individual government departments, such as gas board or road board areas. Scholars making comparative studies of the political geography of local government in different countries may find the classification suggested by Alderfer (1964) useful. He distinguishes between the French, English, Soviet and traditional patterns.

> . . . French local government is characterised by centralisation, chain of command, hierarchical structure, executive domination and legislative subordination.
>
> (p. 7)

English local government is characterised by decentralisation, legislative dominance, co-option through the committee system, multipurpose activity, and voluntary citizen participation.

(p. 10)

. . . Soviet local government is characterised by Communist party control under the name of democratic centralism, single-candidate elections, hierarchical chain of command, and broad scope of governmental powers to local councils.

(p. 14)

. . . they [traditional residues] are inclined to be simple in political structure, which is backed up by complex sociological mores and originally tended in the direction of democracy and freedom.

(p. 16)

This section can be usefully concluded by reference to three very useful studies by McColl (1963), Helin (1967), and Sautter (1966). McColl reviewed the changes in the administrative pattern of the People's Republic of China on the basis of information derived from contemporary sources. He demonstrated that the first administrative regions were based on the need to overcome remaining pockets of resistance to the revolution and the need to establish effective military government. In 1952 government in these regions was made uniform as a preliminary to the inauguration of the first five year plan. However, after only two years this regional system was exchanged for an arrangement of provinces; McColl (1963, p. 58) believes that this change was designed to halt growing separatist tendencies. In 1955 in an effort to reduce problems of dissidence some minorities were constituted into autonomous areas.

Helin published a very detailed paper recording the changes in the administrative boundaries and system of Rumania between 1918 and 1960. He convincingly relates the system adopted at any particular time to the views of the government of the day; the excessive fragmentation created by the Averescu Government in 1919 was designed to keep the power centralised, and it contrasted with the seven regions of the National-Peasant Government (1928–30) which sought to give expression to the different cultures and aspirations of people in various parts of the state. These two papers provide further evidence that administrative changes are a common feature of authoritarian governments, and it is interesting that in Rumania since 1945 there have been significant changes in administrative boundaries even though all the governments have been communist. Helin attributes these changes partly to the changes in the leadership of the Central Government.

The only aspect of this subject not adequately considered by these papers is the effect on the landscape of the boundary changes; it is probable that this omission was enforced by limitations of space.

Sautter's study of the Republics of Congo (Brazzaville) and Gabon includes a chapter dealing with the evolution of the administrative map in both these areas. The information accumulated and mapped is very useful, but there is regrettably only slight reference to the reasons why boundary changes were made from time to time, and no discussion of the effects of these boundary changes. Figure 3 was prepared on the basis of Sautter's maps (vol. 1, pp. 180–3) and the 1966 changes, to show the length of time which present boundaries have endured as elements of the landscape. This method (Prescott, 1967, p. 157) has attracted little attention from geographers, it is clearly most useful where the boundaries all belong to one political level; no satisfactory arrangement has been suggested where boundaries are transferred from one level such as internal to another, such as international. The map of the Congolese boundaries shows considerable variations and prompts questions about the reason for the longevity of certain lines. The existence of rivers and the distribution of tribes clearly offers explanation of part of the pattern, but would be insufficient to explain the alignment of some of the more recent boundaries.

Electoral policies

Pounds (1963, p. 212) regards electorates as the most important of special purpose administrative areas. This seems a sound view since they are a vital part of the process by which a government is chosen, whether this involves change or not.

In a review of electoral geography's aims and functions (Prescott, 1959) it was suggested that the main interest of the political geographer lay in using statistics to identify political regions, rather than simply explaining the crucial geographical factors which contributed to a particular pattern. In the papers which have been published since this survey the tendency in all cases has been to explain elections in terms of geographical factors. McGee (1962 and 1965) examined elections in Malaya in 1959 and 1964. In the first paper he bravely made certain predictions about the next election, which, for a variety of factors

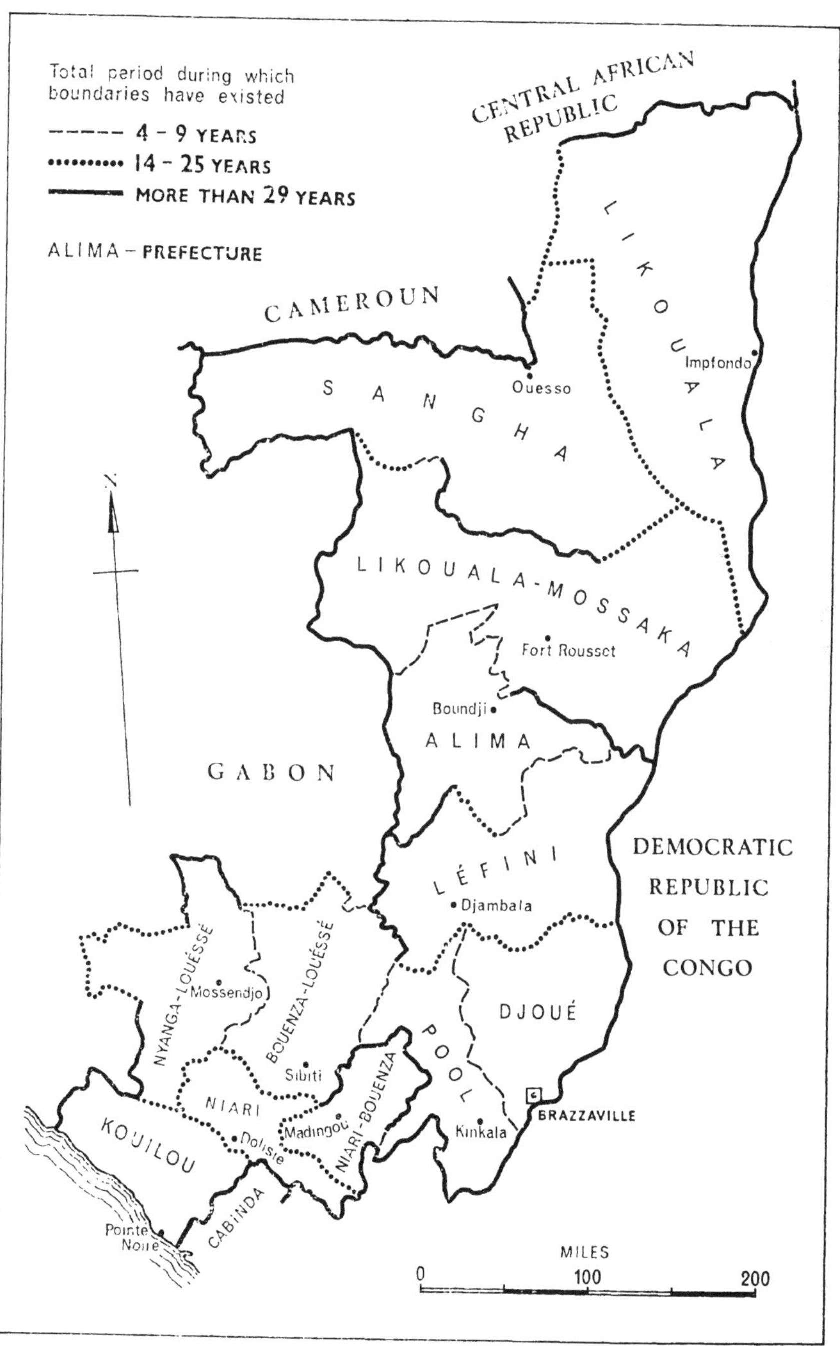

Fig. 3
Boundary persistence in Congo (Brazzaville)

noted in the second paper, were not fulfilled. He was largely concerned with the correspondence between the distribution of racial groups and election returns. Burghardt (1963) identified regions of political support in Burgenland (Austria) during four elections from 1949 until 1959, and explained the geographical factors which helped to explain their distribution. Lewis (1965) examined, by means of a series of maps, the influence of immigration of Negroes into Flint, Michigan, between 1932 and 1962, on the electoral patterns of the area. He demonstrates a very convincing relationship, and points the way to further detailed electoral analyses of small areas by geographers. Roberts and Rumage (1965) made a statistical analysis of the geographical factors which contributed to an explanation of the pattern of votes cast for the Labour Party in 157 towns in England and Wales. They used seven regression models, after eliminating four variables which were duplicating the effects of one or more of the original seven. They concluded that the seven variables statistically explained 81 per cent of the areal variation of Labour votes in the towns studied. The final conclusions that the Labour Party vote was higher in those industrialised towns lying close to coal fields occupied by voters in 'the lower echelons of the social class hierarchy', and lower in those areas where the bulk of the population was classified as part of the 'upper and middle strata of the social class hierarchy' can have surprised no one, least of all Krehbiel who had made a pioneer study of elections in Britain in 1916.

Woolmington (1966, chapters 5, 7 and 8) refers to the results of a referendum and various elections in the north-east of New South Wales in searching for a territorial definition of the New State movement in this area. His analysis of elections is based on 'consistent, clear majority, first preference voting' for the Country Party and the Australian Labour Party, as being the two parties which he identifies as the main supporters and opponents of the New State movement. It is difficult to understand why he eliminates the Democratic Labour Party as a supporter of the idea, on the grounds that its voters may base their allegiance on some other aspect of policy than the New State movement. This argument could logically apply to the other parties concerned. It is also surprising that he feels that electoral boundary changes during the period under review, were unimportant (Woolmington, 1966, p. 78). This may be the case but it would be better to demonstrate the evidence for this opinion.

The paper by Ahmad (1966) is not considered here because it simply restated material treated by earlier students of the subject without adding any new concepts.

It is recognised now that the earlier suggestion (Prescott, 1959) that the main purpose of electoral geography is to help identify political regions within the state is only half the complete study. Geographers have devoted most of their attention to explaining the results of elections and drawing inferences about the political geography of the state. It is just as important to explore the geographical factors which have influenced a government to select one electoral system, rather than another, and to adopt a particular division of the state into electorates when the electoral system requires this. Burghardt (1963) does not discuss the pattern of electorates, nor guarantee that they were unchanged during the four elections; Roberts and Rumage (1965) do not include any discussion on the problems of matching electorates and census districts, although Butler, a leading election analyst, regards this as a critical difficulty.

A study of electoral policy must concentrate firstly on the electoral system selected, together with any conditions which may be attached to it, and secondly, on the territorial division of the state where this is necessary.

For example, in Ceylon the government has abolished the multi-member constituencies, because this was depriving the minorities of representation. Now the electorates for the 145 seats in the lower house are arranged to allow representation by minorities connected by race, religion or other strong ties. If the Governor believes that certain important minorities are inadequately represented after an election, he may nominate up to six members for this purpose. In the Lebanon candidates are elected to represent the whole nation, not individual constituencies, but the Electoral Act of 1960 stipulates that there must be six Christians for every five Moslems in the Chamber of Deputies. Complicated electoral systems are common where a colonial possession is in the process of achieving independence. The 1961 constitution for Rhodesia laid down an electoral system which was designed to lead the country gradually to African majority rule. The Legislative Assembly was composed of sixty-five members, of whom fifty would be elected from constituencies by A Roll voters, and fifteen from electoral districts by B Roll voters. The two voting rolls had common conditions concerning citizenship, age, residence and language; the difference between the two rolls lay in the additional qualifications. These were a combination of income and education, arranged in such a way that as the income requirement was reduced the educational requirements rose. The practical effect was that most Europeans could qualify for the A Roll, while there would be a majority of Africans on the B Roll. Gradually as more and more Africans qualified for the A Roll, more and more of their representatives would be elected, until a point was reached where they would form a majority in the Legislative Assembly. Both the constituencies and the electoral districts were drawn so that any part of the whole country was included in a constituency and an electoral district.

It is clear from the examples given here, and examples given earlier about the representation of Turkish Cypriots in Cyprus, that a government can obtain a desired result through the conditions laid down for the conduct of the election. This generalisation will also apply in the case of referenda. For example, the referendum held in the counties of Buyaga and Bugangazzi in Uganda in November 1964 was restricted to voters who were eligible to vote in the 1962 general election. This disqualified 20,000 Buganda ex-servicemen who had settled in the areas since 1962 and who could have been expected to vote in favour of continued union with Buganda. The referendum resulted in majorities of 1,000 and 7,000 in Buyaga and Bugangazzi respectively in favour of reunion with Bunyoro. There were similar complaints by the Somali Government against the conduct of the referendum in French Somaliland (now the Territory of the Afars and Issas) in March 1967; it was alleged that Somali voters had been arrested and expelled before the election.

Turning now to the geographical factors involved in any division of the state into electorates there are many reports by electoral commissions which give considerable insight into the process. For example in 1967 the report on the electoral redistribution of Papua and New Guinea was published, together with a series of maps showing the limits of each electorate. The Commission was charged to create sixty-nine open electorates, each with a population of about 30,000, and fifteen regional electorates. Both systems were to cover the whole area and it was stipulated that each regional electorate should be superimposed over the open electorates so that they coincided with District boundaries. The Commission discovered certain problems during their work. First the latest population figures of 2,183,036 gave an average population in each open electorate of 31,638. In fact no permissible deviations had been stated and eventually the largest electorate, Kainantu, contained 45,327 while the smallest electorate, Manus, contained 20,647. Second, the Commission discovered that the eighteen Districts had two boundaries; the official gazetted boundary and the unofficial boundary of the area actually administered. This situation is common in colonial territories, and in South America caused considerable problems about national boundaries when the area obtained its independence. Fifteen of the eighteen Districts administered areas outside their gazetted limits, usually because a group of people were separated from the *de jure* administrative centre by some major physical obstacle. The *de facto* administrative limits were selected as the boundaries to which the electoral limits should correspond. The Commission also considered the community of interest, local government boundaries, the planned movement of people into existing Districts, means of communications, physical features, census divisions, and existing electoral boundaries. When their first plan was outlined there were twenty objections within the allowed period. Nine of these were accepted and eleven were rejected, on grounds that they would split local government areas or Districts, or because of tribal enmities, or because the complaint was based on personal grounds.

The Victorian Government (1965) issued a report concerning redivision of the southern area of the state into electoral provinces for the Legislative Council and of the whole state into electoral districts for the Legislative Assembly. In the case of the reallocation of provinces the Commission was required to aim at an approximate population of 117,000 in each case, with a margin of 10 per cent above and below that figure. In fact the maximum deviation was 2·68 per cent above the figure in Melbourne Province and 4·81 per cent below the figure in Melbourne West Province. In the electoral districts the Commissioners were given three sets of target figures. The forty-four metropolitan seats, defined as the Port Phillip Area, had a quota of 25,000 electors each; the eight urban seats associated with the provincial centres of Ballarat, Bendigo, Geelong and the Latrobe Valley had a quota of 22,250 electors; the twenty-one rural seats had a quota of 18,200 voters. The Commission was also instructed by the Act appointing them to give due consideration to four factors in determining the boundaries: the likelihood of changes in the number of electors in the foreseeable future, community or diversity of interest, means of communication, and physical features. In constructing the metropolitan areas, the Commissioners noted that they had used rivers and creeks and major thoroughfares as boundaries, so that people would be aware of the limits, and to reduce the inconvenience of having to cross such obstacles to vote. In forming the eight electorates in the four provincial centres, the Commissioners found that in every case, except Geelong, it was necessary to add considerable rural areas in order to reach the number of voters required; in such cases this factor took precedence over questions of diversity of interest and access. The lower number of electors in the rural districts gave an advantage to the Country Party, which represents country interests. This position agrees with the review which Pounds (1963) expressed.

> The practice of weighting electoral districts in favour of the rural voter is one of the most persistent of gerrymander devices, and one most difficult to eradicate.
>
> (p. 214)

Pounds (1963, pp. 213–15) gives a number of very clear examples of gerrymanders which have involved the careful delineation of boundaries so that the governing party wins seats with small majorities, while its opponents win a smaller number of seats with large majorities.

Conclusion

Administrative policies form a significant sector of state policies and should be studied by political geographers interested in this field. The importance of such analysis is related to the role of administrative policies in shaping the landscape. This research will demand an examination of relevant constitutional and legal documents, but real understanding can only be achieved in this way. Part of this field has already been developed, notably the establishment of local and federal boundaries; the formation of electorates and policies relating to land ownership and use are two aspects which require further attention. The student of administrative policies should seek to answer four questions.

> What geographical factors, if any, influenced the government in its selection of this particular administrative system?
>
> What geographical factors, if any, influenced the government during the division of the states into specific areas for specific purposes?

To which areas of the state do these administrative policies apply?

What are the geographical consequences, if any, of the application of these policies to these areas?

When the political geographer can answer these questions the subject will have been advanced, and he will be in a position to help in any reconstruction of the geographical aspects of administration.

REFERENCES

AHMAD, H., 1966, 'Election data analysis as a tool of research in political geography', *Pakistan Geographical Rev.*, **21**, pp. 34–40.

ALDERFER, H. F., 1964, *Local government in developing countries*, New York.

BROOKFIELD, H. C., 1957, 'Some geographic implications of the apartheid and partnership policies of southern Africa', *Transactions*, Institute of British Geographers, **23**, pp. 225–47.

BUCKHOLTS, P., 1966, *Political geography*, New York.

BURGHARDT, A., 1963, 'Regions of party support in Burgenland (Austria)', *Canadian Geographer*, **7**, pp. 91–8.

DICKINSON, R. E., 1960, *City, region and regionalism*, London.

FAWCETT, C. B., 1961, *Provinces of England: a study of some geographical aspects of devolution*, revised by W. G. East and S. W. Wooldridge, London.

FESLER, J. W., 1949, *Area and administration*, Alabama.

FISHER, C. A., 1963, 'The Malaysian Federation, Indonesia and the Philippines: a study in political geography', *Geogr. J.*, **129**, pp. 311–28.

GILBERT, E. W., 1939, 'Practical regionalism in England and Wales', *Geogr. J.*, **94**, pp. 24–44.

GOVERNMENT OF VICTORIA, 1965, *Report by the Commissioners appointed for the purpose of the proposed redivision of the 'Southern Area' into Electoral Provinces for the Legislative Council and of the whole of Victoria into Electoral Districts for the Legislative Assembly*, Melbourne.

HAILEY, Lord, 1957, *An African survey: revised 1956*, London.

HALL, J. W., 1966, *Government and local power in Japan 500–1700: a study based on Bizen Province*, New Jersey.

HELIN, R. A., 1967, 'The volatile administrative map of Rumania', *Annals*, Association of American Geographers, **57**, pp. 481–502.

HINDEN, R. (Ed.), 1950, *Local government and the colonies*, London.

HOLMES, J. M., 1944, *The geographical basis of government: specially applied to New South Wales*, Sydney.

JONES, N. S. C., 1965, 'The decolonisation of the White Highlands of Kenya', *Geogr. J.*, **131**, pp. 186–201.

KREHBIEL, E., 1916, 'Geographic influences in British elections', *Geogr. Rev.*, **2**, pp. 419–32.

LEWIS, P. C., 1965, 'Impact of Negro migrations on the electoral geography of Flint, Michigan, 1932–62: a cartographic analysis', *Annals*, Association of American Geographers, **55**, pp. 1–25.

LOCAL GOVERNMENT, 1956, *Areas and status of local authorities in England and Wales*, Cmd. 9831, H.M.S.O., London.

LIPMAN, V. D., 1949, *Local government areas* 1834–45, Oxford.

MCCOLL, R. W., 1963, 'Development of supra-provincial administrative regions in Communist China 1949–1960', *Pacific Viewpoint*, **4**, pp. 53–64.

MCGEE, T. G., 1962, 'The Malayan elections of 1959: a study in electoral geography', *J. of Tropical Geography*, **16**, pp. 72–99.

MCGEE, T. G., 1965, 'The Malayan parliamentary elections 1964', *Pacific Viewpoint*, **6**, pp. 96–101.

MORGAN, W. T. W., 1963, 'The White Highlands of Kenya', *Geogr. J.*, **129**, pp. 140–55.

NELSON, H. J., 1952, 'The Vernon area of California: a study of the political factor in urban geography', *Annals*, Association of American Geographers, **42**, pp. 177–91.

NICHOLSON, N. L., 1954, *The boundaries of Canada, its Provinces and Territories*, Department of Mines and Technical Surveys, Geographical Branch, Memoir 2, Ottawa.

PAPUA AND NEW GUINEA ELECTORAL ORDINANCE 1963–7, *Report by the Committee appointed for the purpose of redistributing the territory of Papua and New Guinea into electorates*, Canberra.

POUNDS, N. J. G., 1963, *Political geography*, New York.

PRESCOTT, J. R. V., 1958, 'The geographical basis of the Nigerian Federation', *Nigerian Geogr. J.*, **2**, pp. 1–13.

PRESCOTT, J. R. V., 1959, 'The function and method of electoral geography', *Annals*, Association of American Geographers, **49**, pp. 296–304.

PRESCOTT, J. R. V., 1962, 'The geographical basis of Kenya's political problems', *Australian Outlook*, **16**, pp. 270–82.

PRESCOTT, J. R. V., 1967, *The geography of frontiers and boundaries*, London.

RHODESIA AND NYASALAND, FEDERATION OF, 1961, *Southern Rhodesia (Constitution)* Order in council 2314 of 1961, Salisbury.

RHODESIAN GOVERNMENT, 1965, *Constitution of Rhodesia*, 1965, Salisbury.

ROBERTS, M. C., and Rumage, K. W., 1965, 'The spatial variations in urban left-wing voting in England and Wales in 1951', *Annals*, Association of American Geographers, **55**, pp. 161–78.

ROBINSON, K. W., 1961, 'Sixty years of federation in Australia', *Geogr. Rev.*, **51**, pp. 1–20.

ROYAL COMMISSION ON LOCAL GOVERNMENT IN GREATER LONDON 1957–60, 1960, Cmd. 1164, H.M.S.O., London.

SAUTTER, G., 1966, *De l'Atlantique au fleuve Congo: une géographie du sous-peuplement*, Paris.

STEINER, K., 1965, *Local government in Japan*, Stanford.

TAYLOR, E. G. R., 1942, 'Discussion on the geographical aspects of regional planning', *Geogr. J.*, **99**, pp. 61–80.

ULLMAN, E. L., 1939, 'The eastern Rhode Island–Massachusetts boundary zone', *Geogr. Rev.*, **29**, pp. 291–302.

WHEARE, K. C., 1951, *Federal government*, London.

WHITTLESEY, D., 1944, *The earth and the state*, New York.

WOOLMINGTON, E. R., 1966, *A spatial approach to the measurement of support for the separatist movement in northern New South Wales*, University of New England, Monograph series No. 2, Armidale.

5

POLICIES FOR THE DEVELOPMENT OF THE STATE

Recent studies in agricultural geography hardly accentuate sufficiently the government factor in modern agriculture in highly organised states.

(Large, 1957, p. 365)

National governments influence foreign trade in numerous ways, one of the most important being by legislative controls in the form of tariffs, quotas and embargoes . . . The power of such political factors cannot be exaggerated.

(Alexander, 1963, pp. 510–11)

Political control and national policies are highly important factors in the geography of commodity production.

(Highsmith and Jensen, 1963, p. 2)

Any discussion of the development of rural settlement in Australia cannot ignore the attitude taken by Colonial and State governments to the question of land tenure.

(Camm, 1967, p. 263)

These quotations show that geographers who are concerned with general economic geography, special aspects of that subject, and historical geography are aware of the significance of a government's role in influencing the development of a state, although this awareness has not resulted always in a satisfactory treatment of this material. For example, in three chapters on mining where the influence of governments is marked, Alexander only mentions the political factor twice; once dealing with the way political boundaries cross some mineralised regions, and once when the concept of conservation is mentioned. But political geographers cannot adopt a superior position in this matter for a number of them also fail to give this aspect of the subject suitable consideration. Pounds (1963) and Buckholts (1966) make reference to government economic policies in their general treatments of political geography, but de Blij (1967) manages to consider industrialisation, 'the cornerstone of power', without reference to the policies by which states achieve this development. By the same token Brodie and Doherty, writing on the importance of economic factors in political geography (Weigert, 1957, Pt. 3), make no significant consideration of the way in which political factors are important in the economy of the state, or the way in which government policies may modify economic factors. Alexander (1957), in his study of world political patterns, surveys the economy of Australia, and many other countries, without reference to the fiscal policies which buttress their exports or shield their manufacturing industries.

The policies which are considered in this chapter have three main aims. First there are those which will attempt to increase existing production; second there are those which will stimulate new forms of production; and third, there are those which will simply aim at avoiding any loss of production by maintaining present levels of output. It is perfectly possible for all these policies to coexist in respect of different industries at the same time. For example, during November 1967, the Australian Government announced the allocation of further funds for the Ord Scheme, in Western Australia, to allow the extension of the cultivable area, and approved the prices arranged for iron-ore sales, totalling forty million tons, to Japan by Hamersley Iron Ore Proprietary Limited; both policy decisions allowed the increase of existing outputs. The government also launched an examination of the needs of farmers producing exports arising from the devaluation of sterling, so that existing levels of production could be sustained. Finally, at the end of the month, it was announced that the government was giving serious consideration to the construction of a railway from Duchess, in north-west Queensland, to Bourke, in northern New South Wales, in order to facilitate the development of rich phosphate deposits found in the area of Duchess by Broken Hill South Limited and International Minerals and Chemicals Limited; if this plan comes to fruition it will significantly reduce Australia's import bill for phosphates, which cost the country $A8 millions in foreign exchange in 1966.

It is always possible that the development policies implemented by governments may have a basic political motive, but this will not invalidate their analysis. For example, nationalisation may be undertaken by the governments of some under-developed countries, in order to silence allegations of neo-colonialism by certain sections of the population. Alternatively, aid programmes may have a smaller component of altruism than supposed, because the state giving the aid is seeking political advantages from the state receiving the aid. Lastly, many governments have been charged with timing budget concessions and regional investment according to their electoral needs at any particular time; this allegation was made against the Australian Government when it announced the further financing of the Ord River Scheme shortly before the Senate election in November 1967.

The political geographer, indeed any geographer, who studies a government's development policies is interested in the same three facets which were considered in respect of defence policies. First, the geographical factors which influenced the formation of such policies; which includes consideration of their timing as well as motive; second, it is necessary to identify the geographical circumstances which influence the operation of the policies; and lastly the study must be completed by recognising the geographical results of policy implementation.

The immediate problem concerns the organisation of the available material. As one begins to collect references dealing with the role of government in development it becomes clear that there is a welter of material which ranges from fiscal policies, such as devaluation, at one end of the scale, which at first sight are remote from geography, to the construction of a railway line at the other end of the scale, where the influence of geography is immediately apparent, and where the geographical effects may soon be observed. Clearly no single system will satisfy geographers in every field, but providing the political, economic and historical geographers specify their methods of organisation, it should be possible to collate the information yielded by the various branches to the general benefit of the whole subject. Highsmith and Jensen (1963) consider government policies which affect various types of production, so that the relevant material appears under sections dealing, for example, with fishing, sugar production and manufacturing. This approach is also characteristic of the work by Pounds (1963) who considers government policies in respect of foreign trade, and Buckholts (1966) who considers government action in respect of commercial agriculture, energy and minerals, and industry. This is a perfectly legitimate system for political geographers and many will continue to use it, but since it has been suggested that they should be more concerned with the actions of particular governments, it is also appropriate to classify the material according to the action taken by governments. This is especially true since the geographical factors which influence policy decisions must enter through perception by the government, and since the consequences of policy acts can only follow when the government has implemented a particular policy. The government therefore occupies the pivot position in this study, and connects the work of those who consider the role of geography in policy formation and those who consider the role of policy in the alteration of geographical features. One of the advantages of this particular system is that it will take account of those policies which have no geographical roots but which still have geographical effects on the landscape.

The need to focus on the action of specific governments, which was justified in the opening chapter, is just as relevant in respect of development policies as it was in the case of defence policies. This point is clearly made by Anderson (1967) in his study of the relationships between politics and economic change in Latin America.

> In one country or another, virtually all the techniques and tools of state economic activity which were available to the European and North American statesmen of the age were tested in the Latin American environment. Tariff policies were manipulated and revised as protectionists and free traders succeeded one another in decision-making authority. Schemes to attract immigration, colonisation, and enterprise through liberal concessions of public lands and taxation advantages appear early in the records of many nations. *Latifundia* concentration itself was often an unintentional by-product of early efforts to increase productivity by placing public lands, or Church Lands, in the hands of private proprietors.
>
> (Anderson, 1967, p. 19)

In addition to showing how different governments in different countries adopted new policies which sometimes gave an irregular rate to economic advance, Anderson (1967, pp. 218–24) demonstrates that similar economic problems, occasioned by the world depression of the early thirties, produced different responses by governments in various states. For example, while Colombia,

Venezuela and Bolivia liberalised their policies to meet the exigencies of the depression, Nicaragua, Honduras and Guatemala adopted restrictive, conservative policies to overcome the problems posed.

Having decided in this study to organise the material by policy, rather than by geographical factors, such as position and size, which affect policy decisions, or by activities, such as mining and agriculture, which are affected by policy decisions, it still remains to select those qualities of policies which are considered critical. The two most meaningful qualities seem to be intention and the extent to which policies are decided by a single government or statutory authority. In the first stage of classification it is sufficient to distinguish two main types of motives; those which are aimed to improve the general economic condition of the whole country, and those which are designed to help a particular region of the state or a particular sector of the economy. The distinction between the number of governments involved is essentially between those cases where one government makes the decision and those where more than one government is involved; this will determine whether it is necessary to construct one or more geographical views. There are, of course, cases where a government has sole responsibility for making a decision, but where another government which detects an area of interest may make its opinions known. In this case the opinion of the second government will be considered as one of the factors which are taken into account by the responsible government. The multilateral policies will be considered to be those where more than one government has to make the decision and adhere to the formal agreement.

With this initial simple division of motive and number of governments involved it is possible to construct four groups of policies. First there are the *general unilateral* policies by which the government of a state seeks to promote the general welfare and economic development of the country. Policies belonging to this group include levels of taxation, bank rates, exchange rates, wage-fixing, and attitude to private enterprise. Second, there are the *general multilateral* policies where the government joins with another in seeking conditions which will encourage the overall development of the state. The formation of customs unions and association with international agencies are representative of this group. Third there are the *specific unilateral* policies where the government makes its own decision on measures to assist the development of one region of the state or one sector of the economy. There are many examples of such policies such as reduced freight rates for distant producers, export subsidies for handicapped industries, and encouragement of decentralisation. Lastly, there are the *specific multilateral* policies where the aid of at least one more government is sought in fostering the development of part of the state or part of the economy. Detailed multilateral trade agreements and certain aid contracts are obvious examples of this type of policy.

Five points should be made before starting this detailed examination. First, it is important to stress that this primary division is based on motive and not on subject of policy or the method by which policy is implemented. For example, policies dealing with immigration may be *general unilateral* policies when a state prohibits the immigration of certain racial groups, which it feels may cause friction within the state or compete unfairly with existing workers; on the other hand, policies to attract labour may result in a multilateral agreement with a number of other states from which the immigrants are drawn, such policies would be *general multilateral* policies. Turning to the method by which policies are implemented, it is clear that the means available to a state are limited. Taxation is one common technique, but this can be applied generally so that it will have an overall effect on the development of the country, or selectively, so that people in one part of the state or people employed in one or more particular industries are favoured or handicapped. Further, double taxation agreements with other countries may be aimed at attracting much-needed capital investment. Thus taxation policies will have to be discussed under the headings of *general unilateral*, *general multilateral*, and *specific unilateral* policies.

Second, it is important for geographers involved in this enquiry to understand the processes by which policies are applied. There is no escape from the need to understand thoroughly the operation of a tariff schedule, an immigration law or a change in currency exchange rates, or from the need to distinguish a tax which is simply designed to raise revenue from a tax which is designed to promote or discourage a certain activity; geographers need to identify those policies where the government can *enforce* a certain decision, for example by the issue of licenses for the construction of a factory, and those where the government *hopes* that fiscal measures will encourage persons to act in the desired way. This is not a novel suggestion; there are the precedents of economic geographers studying the works of economists and electoral geographers being involved with the mechanics of voting in order to give their work greater reality. The important point is to remember that such study is only a means to a better comprehension of the geographical effects of such policies, and a clearer understanding why there may be discrepancies between intention and result.

Third, it is useful to recall that normal economic principles are likely to be much more relevant to policies for the development of the state than they are to policies of defence. The earlier discussion showed that in many cases, during times of national need, economic principles were abandoned; it is possible that in certain cases governments will institute development policies which seem at variance with sound business practices, but more often governments will adopt policies which will yield the maximum profit, however that is measured, for the state as a whole.

Fourth, in any consideration of development policies by governments there will always be a basic division between authoritarian systems, where the government exercises complete control over every aspect of the economic life of the state, and the democratic systems, where the government control will vary both in the extent to which it is involved in the economy and degree of control which it exerts over certain sections of the state's economic life. The very wide range of government involvement within the democracies makes it very difficult to make generalisations about the relationships of policy and geography, and underlines the need for geographers to look closely at individual cases in which they are interested.

Fifth, general economic plans present difficulties to this classification of material, because although different sections of the plan fit into different categories, the plan must be viewed as a whole. It is therefore proposed to consider parts of national plans within the appropriate section, and then in conclusion review some of the geographic literature dealing with economic plans for individual countries.

General unilateral policies

The policies considered in this section are those which are decided by a single government to promote the general economic welfare of the state. It has already been noted that authoritarian governments will exercise complete control over economic development, while democratic governments have to rely on a variety of policies to encourage progress in an orderly manner. Such policies will include measures to control the *availability of money*, such as the bank rate, government expenditure, capital inflow, taxes and exchange rates, the *price and availability of commodities* through basic wage policies, pegged prices and restrictive trade regulations, and the *availability of labour* through immigration policies.

The fiscal policies have not received much attention from geographers and this is understandable because they seem remote from geographical factors in many cases. However, the conditions which promote devaluation or a change in the bank rate may sometimes be found to have part of their source in geographical facts, such as drought or optimum growing conditions which produce gluts of tropical crops, or changed patterns of production because of the policies of other states. The geographical effects of these sophisticated fiscal policies are more readily recognisable. The effects of economic mismanagement are found in unemployment, reduced levels of trading, absence of the establishment of new projects, and the abandonment of certain established industries or at least contraction of output. By the same token sound financial measures will produce demonstrable geographical effects such as reduced pools of unemployment, the establishment of new industries, expanded production, and the construction of major new government projects such as dams and hydro-electric stations.

Each geographer engaged in research which touches on these processes will have to discover for himself the significance, if any, of the fiscal policies operating at the time to which his study refers. The information will be found partially in the government legislation and partially in the textbooks about economics, which will deal with concepts of inflation, bank rate, devaluation and foreign exchange control. It is not appropriate, in a geography textbook, to outline the meanings of terms which have been culled from economists; it seems more important to suggest some of the features of such policies which may be of interest to geographers studying certain states. It is therefore proposed to look at some of the more important types of policies as they have been applied in certain states.

First, an examination of fiscal policies in South Africa during the period 1961–6 will show how there is a unity about such policies which are designed to support and complement each other.

Following the events of Sharpeville in 1960 the South African economy faced a crisis of confidence which revealed itself in a steady decline in foreign exchange reserves. In March 1960 these reserves stood at R250 million, by June 1961 they had fallen to R175 million, the lowest point in recent years. Unemployment began to increase as there were cuts in the output of motor vehicles, buildings and consumer products. The flight of capital was stopped by exchange controls introduced in mid-1961, and the economy was bolstered by increased government expenditure on public works. For example, the decision to proceed with the Orange River project was taken at this time, and government and public authority spending rose by R103 million compared with an annual increase of R33 million during the previous four years. Money was made cheaper to borrow by reducing the bank rate from 5 to 4 per cent. Exchange control also resulted in a decline in imports from a level of $1,556 million in 1960 to $1,410 million in 1961. Commercial banks and building societies reduced their overdraft and mortgage rates and the volume of building began to increase. In the 1962 budget the government introduced tax incentives for exporters which undoubtedly contributed to the increase of $66 million in the value of exports during 1962–3. Unemployment declined and some labour shortages began to appear. Late in 1962 the bank rate was further lowered to 3½ per cent and in the following year a rebate of 5 per cent on income tax and a reduction in the price of petrol made more money available and reduced some costs. The cumulative effect of all these measures was a rapid growth in gross national product, which rose by 16 per cent in the two years ending June 1964.

Clearly geography played little role in the initiation of these economic policies, which resulted largely from political considerations, but the geographical effects of these policies were significant to the South African landscape. City skylines changed as new buildings replaced structures which had become too small for the value of the site they occupied; the establishment of the Caltex oil refinery, Hoechst's polyester-fibre factory and Fison's fertiliser plant stimulated the rapid development of Milnerton; Phalaborwa, in the north-east Transvaal, changed from being a small mining town based on phosphate extraction to a major mining area producing significant supplies of copper; work was started on the Hendrik Verwoerd dam, which, when completed, will be the major structure in the Orange River irrigation and power scheme.

By mid-1964 signs of inflation were appearing and during the following year they became clearer as labour shortages developed and prices began to rise. The situation was worsened by a severe drought which cut farm exports just at the time when imports were beginning to rise sharply. The government was now faced with the opposite problem to 1961; now it was necessary to damp down the rate of expansion and the first step was to make money dearer so that capital developments would be discouraged in some cases. The bank rate was raised and so were taxes while state spending was reduced. The South African Government is more sensitive to signs of inflation than many other governments, because of the adverse influence rising prices have on the profitability of gold mining. Since the official price of gold is fixed, any process which increases the costs of mining reduces the profit margin of this industry which is still the most important in South Africa. Between 1961 and 1965 the gold industry provided 40 per cent of the country's exports by value, spent R450 million on extensions, paid R943 millions in wages and salaries, purchased stores worth R1,244 million, and yielded R563 million in taxes to the treasury (*Financial Mail*, 14 July 1967, p. 35). In the same period wages rose by 15 per cent, but the full effect of this development was offset by the growing importance of mines in the Orange Free State and the western Transvaal where ore grades were richer, and will continue to be richer for the next three years. Thereafter, grades of ore in existing mines are likely to decline. During the period 1961–5 it is estimated that ore worth R1,000 million was passed over because it was not economic to process. Unfortunately the industry is unable to rationalise its labour situation in order to cut costs because of race policies which prevent the most economic use of Bantu workers.

In 1966–7 measures were continued to ensure deflation; these included an increase in the bank rate to 6 per cent; a reduction in government spending through the postponement of such projects as the aluminium smelting plant at Richards Bay; price control on selected vital commodities; and greater control on the availability of money for capital development. This last point illustrates the need to be aware that because a government announces a policy it does not follow that the policy will be effective. In order to circumvent the shortage of money many South African firms were borrowing from each other—the so-called grey market; the government acted to stop this in October 1967 (*Financial Gazette*, 6 October 1967).

In this review attention has been focussed on the general policies of the South African Government and a number of points can be made which are of interest to geographers. First the government has exerted a considerable influence on the economy of the country and any geographer studying the economic geography of South Africa must take this into account. This influence has been exerted through financial policies, such as exchange control, and also through government spending on a variety of major products. The significance of geographical factors was not apparent when the current cycle began, but as the cycle progressed geographical factors became important in some measure, both because of the drought which affected agricultural output in certain areas, and because of the nature of the gold industry and its vulnerability to inflation. Second, the geographical effects of such policies have been considerable, whether measured by the establishment of new industries, or by new levels of trade, or by differences in regional prosperity. It must be remembered that these policies were supplemented by specific unilateral policies which helped industries such as goldmining or areas such as the Bantustan borderlands, and by multilateral agreements which improved the country's trade situation.

Turning now to some individual fiscal policies, the first measure is the process of devaluation. Devaluation means that a government decides to reduce the value of the national currency against the values of the currencies of other states: usually the devaluation is expressed as a percentage and the new rates compared in terms of sterling and American dollars. In 1966 India devalued the rupee by 36·5 per cent and in 1967 Britain devalued sterling by 14·3 per cent. In each case a variety of factors were blamed including dock strikes, the Middle East war and the activity of foreign speculators in the case of Britain, and drought and the war with Pakistan in the case of India. But each government stressed that one of the basic causes was declining export revenue and increasing import bills.

> The tendency for our exports to be priced out in foreign markets has been in evidence for some time and the Government had during the last few years taken a number of steps in an attempt to remedy this situation. . . . Subsidies on exports had to be increased periodically and even traditional exports such as tea and jute now require support . . . Despite all such assistance the rising trend in exports has not been maintained and there has actually been a slight fall in our exports in 1965–6. In regard to imports also despite progressive increase in import duties the imported goods still continue to command a premium as Indian prices of comparable goods are well above world prices.
>
> (Keesing's Archives, 1966, p. 21505)

The benefits which devaluation are intended to bring are as follows. Exports now become more competitive in price and larger quantities should be sold. Imports become more expensive and therefore there should be an incentive for domestic industries to produce substitute imports and employ more people. Theoretically this means that exports rise and imports fall at least narrowing any trade deficit and perhaps creating a surplus. Clearly this is a cumulative effect and any geographer studying the geographical consequences would have to look at individual cases. In some cases domestic producers will still be unable to compete with imports even at the new rate of exchange. This is true, for example, of transistor radios from Japan. In some other industries price is less important than performance, as in the case of electronic equipment. In other cases, such as food and chemicals, price is generally the determining factor, and providing the state can produce these items from domestic sources it has a considerable advantage through devaluation. If on the other hand it is necessary to import raw materials for processing and re-export much of the advantage is lost because of the greater cost of the imported raw materials. The British chemical industry which uses large imports of sulphur and crude petroleum falls into this last category (*The Economist*, 25 November 1967, p. 881).

Devaluation sometimes fails to produce the desired result as Ferrer (1967) noted in the case of Argentine. Devaluation occurred in 1955 and 1959 but this failed to stimulate agricultural output for three main reasons. First, the organisation of farming on the basis of tenants and large estates discouraged capital investment which was necessary to raise yields. Second, rising costs offset the gain that might have accrued from increased production; and third, monopolistic practices in the meat industry kept export prices depressed and limited expansion of the industry (Ferrer 1967, pp. 188–9). It is usual for devaluation to be accompanied by a number of other supporting measures. For example, when the Indian Government devalued the rupee they simultaneously ended the subsidies on export industries which had been struggling, imposed export duties on traditional export industries such as jute, tea and cotton, and fixed the price of fertilisers and kerosene to avoid any marked rise in prices. The British Government accompanied devaluation with a rise in the bank rate to 8 per cent, a reinforcement of the voluntary incomes and prices policy, cuts in government spending and more stringent conditions of hire purchase.

The success of the devaluation measures will also depend on the reaction of other countries. Clearly if all other countries devalued by the same amount the advantage would be lost; the extent to which devaluation by one country creates a chain reaction depends on the international importance of the currency. When India devalued no other countries followed suit. When Britain devalued fourteen other countries followed almost immediately; these were countries which had a major proportion of their trade with Britain. Fortunately for Britain her major markets, Australia, South Africa, Canada, the United States and most European countries, did not devalue; many of those countries which did devalue were suppliers of raw material which should restrict rises in Britain's import bill.

The change in a country's exchange rate is obviously a significant event which can have an influence on the economic geography in a number of different ways. Daveau (1959) showed how important it was to understand changes in the exchange rate of French and Swiss francs in order to comprehend the economic geography of the Franco-Swiss borderland in the Jura mountains; it is equally important for both economic geographers concerned with a country's trade and industrial and agricultural development, and political geographers trying to understand the geographical significance of a government's action, to realise that devaluation represents a new and vital factor in the economic environment.

The next financial policy worth consideration concerns the government's attitude to foreign investment in the state. This is a question on which economists and politicians hold widely divergent views.

> Too rich an injection of American dollars for the Australian community to assimilate has been shot into the economy since the immediate aftermath of the second world war; too rich an injection of British pounds sterling, also, for the continued health and growth of the economy, and the social and political community, as an *Australian* economy, an Australian community with its own distinctive character.
>
> (Fitzpatrick and Wheelwright, 1965, p. *ix*)

> I would further submit, on the basis of these facts, that Australia's present-day mining industry has been largely built on overseas capital and that the overall benefits of that capital to Australia have been enormous.
>
> (Fisher, 1967, p. 1)

It is not necessary for political geographers to become involved in this argument; what is necessary is that they should identify the attitude of the government concerned and appreciate the geographical consequences of any policies which ensue from that attitude.

A report by the United Nations (1963) which surveys policies regarding overseas investment in ninety countries shows a wide range of restrictions. For example, Mexico allows only minority foreign holdings in a number of industries which are considered strategically important, such as airlines, mining, shipping, fertilisers and rubber; no foreign interests are allowed in the basic petrochemical industry. In France permission must be obtained from the Foreign Investment office before capital can be imported and restrictions have been placed on foreign interests in mining, shipping and pharmaceuticals (Fitzpatrick and Wheelwright, 1965, p. 175). Some writers believe that Japan possesses the most stringent and efficient control over foreign investment; capital allowed in must make a welcome addition to the economy, support the country's foreign exchange position and avoid competition with existing industries.

By contrast Australian Governments have not imposed any statutory regulations on the entry of foreign capital although there is reason to believe that influence is exerted through foreign exchange control. Analysis of available Australian statistics shows the importance of Britain and America as sources of capital, and the growing importance of mining and quarrying, oil exploration, food, drink and tobacco industries and insurance and financial transactions as targets for investment. Fluctuations in the total volume of overseas investments are likely to be the result of actions by other governments and the attractiveness of other countries. For example, in February and April 1965 the American and British Governments respectively enacted measures designed to reduce the outflow of capital to other countries.

The geographical consequences of the Australian Government's attitude towards foreign investment are to be seen in the extension of mineral exploration programmes, the increasing output from new mineral deposits, the construction of railways and ports to service these deposits, such as iron ore at Hamersley and bauxite in the Gove Peninsula, the extension of cotton growing in the Namoi Valley and sorghum cropping in the Northern Territory. Fisher (1967) noted the reverse situation in Mexico.

Mexico is typical of those countries which have impetuously starved themselves of mining capital by over-taxing and harrying the outside investors. According to a report of the International Bank for Reconstruction and Development in 1953: 'Whatever the aim of official policy, its chief effect has been to discourage new investment in the industry. The negative results of this policy are now apparent in the creeping stagnation overtaking the Mexican mining industry.'

(p. 8)

States which have undeveloped resources and a shortage of capital and which have no economic or political fears about foreign investment may set out to attract capital by a variety of policies. These will include freedom to repatriate principal and dividends, tariff concessions on necessary imports and guarantees against nationalisation. Many of the new states of Africa have given such guarantees (Prescott, 1966, p. 303). But favourable conditions are not enough as Ferrer noted in the case of the Argentine.

. . . since the great depression long-term private capital has gone mainly into the exploitation of natural resources such as petroleum and non-ferrous metals in strong demand abroad, and into industrial activities with a sizable and rapidly expanding domestic market. Favourable legal and administrative conditions are not enough; the general economy must be attractive to capital inflow.

(Ferrer, 1967, p. 191)

It has already been noted in the chapter on defence policies that the corrollary to states seeking to attract foreign capital occurs when states use capital to gain control of strategic resources in other countries in order to increase the state's strength.

Guarantees against nationalisation are obvious measures to encourage the investors of other countries, who have a natural fear that their holdings may be rendered worthless by the action of another government. In Africa, during 1966–7, Congo (Brazzaville), Zambia, Senegal, Somalia and Malawi gave guarantees against nationalisation; so did many other African states, but the reliability of their promises must have been compromised by acts of nationalisation which had been taken or planned.

When a political geographer looks at nationalisation there is a need to distinguish those states where the government controls all economic activity from those where government and private enterprise control differing shares of the economy. Total government control of a whole country or a whole industry within a country reveals itself to the geographer through the uniformity of operation, the lack of competition, and the ability of the government to subordinate economic realities to political necessities. This would probably be most apparent in comparisons of railway systems developed under a free-enterprise system, as in America, with the system established under government control, in a country such as Australia.

When attention is focussed on countries where there is a mixture of public and private enterprise it seems important to distinguish acts of nationalisation in countries where there is mainly domestic economic activity, from those where economic activity is mainly in the hands of aliens. The motive in the first case will normally be greater efficiency and rationalisation. The nationalisation of domestic industries, such as Britain's nationalisation of the coal and iron and steel industries, are based on sound economic arguments aiming at rationalisation and increased efficiency. The nationalisation of alien industries usually stem from political motives, which are based on fears of exploitation or manipulation by foreign interests. Since the decolonisation of Africa gathered momentum after 1957, there have been many acts of nationalisation in the continent, and a survey of the two years 1966 and 1967 shows a fair sample of the range of policies. The first major group involves acquiring control over the means of production by nationalising land, factories and businesses; the second major group seeks to provide improved opportunities for the indigenes of a particular state.

Nationalisation of land is mainly designed to make land available for peasants who were previously landless. In Morocco and Tunisia, in 1963 and 1964 respectively, foreign-owned land was confiscated. In Morocco the area involved was 543,000 acres compared with 101,270 acres in Tunisia. The French estimate of alien land confiscated in Tunisia was 750,000 acres, but this included land grants made by French Governments, which were specifically excluded by Tunisia. During 1966 the Moroccan Government distributed 14,800 acres to 501 families, who are grouped into thirteen co-operative farms. Tunisia distributed half the appropriated land to 900 farmers and placed the rest under the supervision of 100 farmers for the benefit of the state. This outright confiscation of land, with protracted discussions afterwards about compensation, contrasts with the scheme in Kenya. The Kenyan land settlement scheme, like the Moroccan and Tunisian actions, seeks to establish a landowning African community in scheduled areas previously owned by Europeans. But in this case the land has been bought from the European farmers and then distributed to Africans in high and low density systems which accord with the agricultural potential of the land. By June 1966, 35,000 African families had been resettled in the former scheduled areas. This reorganisation has altered such features as population distribution, agricultural output, road networks, and patterns of regional exchange. Malek (1966) has discussed some of the geographical consequences of land redistribution in Iran, in a way which suggests fruitful lines of research.

Export industries and commercial establishments are also frequent targets for nationalisation. For example, in May 1966 the Algerian Government nationalised the iron and copper mines at Ouenza-Bou Khadra, the lead and zinc mines at El Abed, Ouarsenis and Sidi Kamber, and the iron mines at Gara Djebilet in the Western Sahara. This last mine lies in territory which is claimed by Morocco. Morocco was also affected by the nationalisation of the El Abed mine, for the lead and zinc deposit straddles the boundary, and the material from both sides is processed in Morocco. The Moroccan miners in the Algerian mine were repatriated during the following month. In October 1967 the Tanzanian Government nationalised thirty-nine sisal estates which produced 60 per cent of Tanzania's output of 220,000 tons in 1965. This action was justified by the Minister for Development, on the grounds that

> . . . the price fall has brought many difficulties to the farmers and the Government has received requests from farmers to help in different spheres.
>
> (*Africa Research Bulletin*, 1967, p. 858)

It is believed that the decision of the Union Minière Company to first lower, then raise, the price of copper without reference to the Congolese Government was partly responsible for the government's act of nationalisation in January 1967. In some cases, where the transfer of ownership does not involve any change in staff, or level of production, or price policy, or change in patterns of marketing, the act of nationalisation will be unimportant to the geographer; but sometimes the change of ownership will involve policy changes which will reveal themselves during the normal course of geographic research.

Looking at the acts designed to increase the opportunities for indigenous citizens, which like nationalisation is an act of economic nationalism, policies vary from the exclusion of aliens from certain activities to the enforced substitution of indigenes for expatriate staff. For example, during 1967, the Sudanese and Libyan Governments reserved the import-export business for nationals. Later in the same year the Kenyan Government restricted non-urban trade and trade in specific commodities to Kenyan nationals, with the proviso that some aliens may be allowed to continue trading if they obtain 'express authority'. This policy is linked with stringent immigration conditions which came into force in December 1967. These regulations abolish the temporary working permits and lists twelve categories of non-citizens who may be given an entry permit and the right to work, providing they undertake to train a Kenyan to take their place within a specified time. The largest alien population is drawn from India and Pakistan and numbers 185,000, of whom 40,000 are citizens and another 8,600 have been naturalised. It will be interesting to see how the number of Asians in Kenya varies as a consequence of this Act, and the extent to which the rural retail trade, traditionally their preserve, becomes a predominantly African occupation. The Liberian Government has declared certain occupations as being reserved for nationals; the industries specified in this way include the sale of meat and bread, dry cleaning and public transport. The Sudanese Government increased the opportunities for their countrymen, in February 1967, by compelling all businesses to sell half their shares to Sudanese immediately and to sell the other half to Sudanese firms within five years. The Governments of Uganda and Kenya have expressed concern at the slow rate at which Africans are securing executive positions in private firms, compared with the much faster rate in the public service, and pressure is being brought to force firms to accept more Africans in managerial positions.

Clearly the implementation of these policy decisions will not always create changes in the economic or political geography of the state, but students must be aware of the possibility and look at each case separately. It is also important to look at the effects which may be produced in other countries than those where the policies operate. For example, it is considered that Tanzania's policies of nationalisation are likely to improve the chances that investors will prefer to place funds in Uganda and Kenya, especially since these countries have made statements denying any intention of carrying out such policies. The confiscation of Belgian interests by the government in Kinshasa, and of land belonging to Frenchmen by the Tunisian Government, is known to have caused friction between the governments concerned, and, in each case, an emigration of Belgians and Frenchmen, and a reduction in the amount of aid given by the European countries to the African states. It will be interesting to assess the geographical significance of any significant exodus of Asians from Kenya, on the areas in which they resettle.

Having discussed policies which influence the availability of capital and the means of production, it is now necessary to consider policies relating to labour in the state. At this stage attention is focussed on the *general* policies which aim to improve the general situation of labour throughout the state; policies seeking to improve a labour situation in a particular part of the country, or in a particular industry, are considered later. Pounds (1963) and Buckholts (1966) have shown an awareness of the political geographical significance of labour policies, especially as they operate on migrations between states. One of the best theoretical accounts of labour policies is contained in a United Nations' report on *National and International Measures for Full Employment* (1949). The basic aim of governments in respect of labour is to create a condition of full employment related to the resources of the country. Clearly different states will face different problems. The British Government in late 1967 faced the problem of finding work for men made idle by stringent economic policies; the Australian Government since the Second World War has faced the problem of attracting labour so that the undoubted resources of the continent can be adequately developed; the Maltese Government, since independence, and the British administration before that, has been concerned at the disproportion between Malta's resources and population, and has encouraged a measure of emigration from the island.

Unemployment generally stems from one or more of three causes: lack of resources, seasonal fluctuations, and insufficiency and instability of effective demand (United Nations, 1949, p. 11). Some of these causes may arise solely within the country, as in the case of inadequate resources, some may originate within other states, such as declining demand for a basic export; it therefore follows that governments will be able to take unilateral action to improve the situation, but they will also have to cooperate with other states in certain cases. The domestic, unilateral policies which are under consideration at the present can either tackle the problem by seeking to stabilise investment or demand.

The ways of encouraging investment include manipulation of interest rates and government expenditure, which have already been considered. Demand can be stabilised through sensitive taxation, subsidies and direct payments. These are the kinds of policies which will be introduced to soak up unemployment, and geographers should be aware of them, since their success or failure may reveal itself in the occupation structure and regional patterns of employment throughout the state. Pounds (1963, pp. 125–6) has written of the need to identify the characteristic of under-employment in a state, and the existence of this problem in under-developed countries was noted by another United Nations report (1949a). Under-employment is really unemployment disguised by the presence of workers in primary industries who could be withdrawn without any loss in production. Overall economic development and significant emigration are the only policies which can deal with this problem.

States which are short of workers, such as Australia and South Africa, have to develop policies which attract migrants of the desired type. Australia has assisted migrant schemes for permanent settlers; South Africa has similar schemes for Europeans who supply the skilled labour component, and efficient short-term schemes for semi-skilled miners and farmers from countries such as Malawi, Botswana and Lesotho. These countries are in the position of being deficient in resources, and accordingly they encourage these short-term movements which serve the twofold purpose of reducing pressure on resources within the country and earning foreign exchange which is remitted home in significant amounts. Geographers have considered some of the consequences of these migration policies which aim to satisfy a labour shortage. Lee (1966) examined the distribution of Italian migrants in Melbourne; Scott (1954) and Prescott (1959) have considered the patterns of migrant labour in southern Africa.

Specific unilateral policies

While the general policies considered above were aimed at the development of the country's economy as a whole, the specific policies considered in this section are concerned with a particular area of the state or a particular sector of the economy or a particular section of the population. Although emphasis will be placed on policies promoting development, or avoiding regression, it is equally essential that geographers should be aware of those cases where government policies, usually unwittingly, cause a decline in the economic productivity of certain areas or sectors. Clearly the targets of such policies will merge when an area of the state has a marked specialisation in production, as in the coal-mining areas of South Wales during the inter-war period. Political and economic geographers interested in this field will find that other scholars share the burden.

> Economists have long recognised the existence and stubborn persistence of regional dualism at all levels of national development and throughout the historical experience of almost all presently developed countries.
> (Williamson, 1965, p. 3)

Reading through the considerable literature by economists and others concerned with regional inequalities the absence of reference to work by geographers is striking. For example, McCrone (1965, pp. 120–3) discusses the problems of identifying regions without any reference to the enormous amount of work published by geographers on this question. It is possible that this neglect is due partially to the tendency of geographers to be more concerned with the effects of policies rather than their formulation.

There are a number of reasons which may cause a government to take particular interest in the development of one part of the state. First, the region may be relatively depressed. This is a term which needs some qualification as Wilson (1965) and Thomas (1962) have shown. Some would identify depressed areas as those in which *per capita* income, or the rate of economic growth, is less than the national average, or where there is a significant measure of emigration, or where unemployment exceeds the national average. The limits of depressed areas based on these criteria would not coincide, although it is likely that the core areas would be the same in each case. A survey of papers by Bowden (1965), Humphrys (1965), MacLennan (1965) and Parks (1965) dealing with north-east England, South Wales, France and the Atlantic provinces of Canada respectively, shows that a higher than average rate of unemployment and a loss of population through migration are the two most widely accepted characteristics of depressed areas. These qualities can develop in areas for a variety of reasons. The exhaustion of mineral deposits has been responsible for depression in some areas in every continent which lacked alternative means of employment.

> It is in West Durham that the coal-mining industry's problems are most acute. There, seams are generally thin, near the surface, and

lacking in continuity; costs of extraction are high, mechanisation is difficult or uneconomic, and the workable reserves of many pits are approaching exhaustion . . . by the end of this century it seems probable that coalmining will have virtually ceased in the western part of the coalfield.

(Bowden, 1965, p. 21)

For the non-extractive activities decline may be caused through the loss of markets or the loss of profitability. Lost markets may stem from the cyclical activity experienced in industries producing capital goods; such fluctuations create the major problems found in Tyneside and Northern Ireland. Markets can also be lost through the raising of tariff walls by other states, which prevents profitable competition with domestic producers. Writing of the eastern seaboard of Canada, Parks (1965) makes the following point:

The production of end products using the output of resource and primary processing activities as inputs has also been retarded by foreign, principally American, tariffs on fully and chiefly manufactured goods.

(p. 79)

Profitability may diminish in primary production through price decline; thus areas depending on such revenue fare badly during periods when world commodity prices are falling. Lastly, substitution or the introduction of new techniques may cause a loss of markets. Copper-producing countries are very conscious of the dangers associated with a significant increase in copper prices, because of the incentive this gives for the substitution of aluminium and plastics; natural fibre producers demonstrate concern at every new development in the field of artificial fibres; the familiar cry of the dairy farmer against the producer of margarine has been heard in most developed countries; and the technique for extracting sulphur from sour natural gas will show an increasing adverse effect on producers using more traditional methods (Seawall, 1961).

The other method by which a region may be adversely affected is through some natural disaster such as flood, earthquake or drought. Most governments from time to time have to take action to redress the misfortune of a particular area. This is not a subject which has attracted widespread attention from geographers, and much of the attention to floods and drought has been given by geomorphologists and climatologists rather than economic geographers. One of the most interesting studies of natural disasters was made by Freeberne (1962) when he studied the position in China from 1949 to 1961 on the basis of a wide variety of contemporary reports. He gives a detailed account of the effects which were admitted after natural calamities, suggesting that this is probably the minimum level cf significance. Decline in grain output, regional migration, and suspension of industrial production all occurred at some time or other during the period studied; and these interruptions brought responses in the form of government efforts to restore production.

But governments do not only show an interest in specific areas because they are depressed; they also concern themselves with areas which appear to have an unrealised potential for rapid expansion. This potential may be in the form of newly discovered resources. During 1965–7 there was a marked expansion of mining exploration, discovery and exploitation in Western Australia. The new ventures included nickel-mining in the Kambalda area and iron-ore mining at Hamersley; the Governments of Australia and Western Australia have encouraged such developments. The new resource might also be provided through a revaluation of the environment for agricultural production through the provision of water or the development of new strains of plants which can grow in previously unsuitable areas. Attention is also likely to be directed to areas whose output will reduce the need to import certain primary raw materials or manufactured goods. The Australian Government has given some support to ventures which will produce petroleum from Bass Strait and phosphates from Queensland; at present Australia imports the bulk of these requirements. Finally there may be strategic reasons why a government decides to encourage development in a particular region. For example, Hamilton (1964) showed the significance of strategical considerations in the location of Yugoslav's iron and steel industries after 1948:

> . . . the threat of attack from the north and east resulted in the location of some capacity in strategically more-isolated and more-secure areas [in 1949] in mountainous central and southwestern regions (Bosnia and Montenegro) . . . after 1948, development, at Smederevo, was discouraged because of its proximity to the vulnerable northeastern frontier.
>
> (p. 60)

An area may also receive attention from the government if its development is at such a rapid rate that problems of congestion, vulnerability and inefficiency begin to arise. British geographers have made a number of studies of government action designed to reduce the population growth of south-east England through the creation of new towns in other parts of the country; the study by Edwards (1964) provides a useful account of this problem.

When attention is turned to the factors which will adversely affect a sector of the economy, it follows that they will in most cases be identical with those already considered as affecting regions, because a region is influenced through the types of production which are carried on there. Natural disasters will normally affect a region rather than a sector of the economy, although certain epidemics, such as foot and mouth disease, can affect a single activity in widely separated areas, as the 1967 outbreak in Britain showed.

Sections of the state's population will be the subject of state policies if they have certain special characteristics which distinguish them from the rest of the population. Racial and ethnic differences spring to mind, but these have already been considered under the heading of strategic policies designed to avoid or eliminate dissidence. There are other qualities within an ethnically homogeneous population which will prompt government policies. For example, many governments have schemes which help migrants to pass through the transition stage of adjustment as quickly as possible; this is particularly true of countries such as Israel, South Africa and Australia which have vigorous campaigns to attract migrants. Refugees are another category of people who usually receive sympathetic consideration from the government; in some cases, as in the Indian subcontinent and the Middle East these refugees are likely to be concentrated in particular areas, giving the problem a regional characteristic. Lastly, in many countries special efforts are made at the conclusion of a war to help returned servicemen to settle back into civilian life. For example, after the Second World War the War Service Land Settlement Scheme, instituted by the Australian Government, provided for the establishment of returned servicemen on the land; to July 1965, 9,144 farms had been established under this scheme, covering nearly fourteen million acres, principally in New South Wales.

Having identified some of the reasons why areas of land, sectors of the economy, or sections of the population may require special attention by the government, it is necessary to consider the policies by which such government attention is demonstrated. Such policies may have one of a number of aims. In some cases, where an area or sector is depressed the government may seek to restore them to former levels of prosperity; in other cases where this is not possible the policy may serve to cushion the area or sector against the worst effects of the depression while it makes adjustment to new forms and patterns of production. In the case of undeveloped or under-developed areas or sectors the policies will seek to make them as productive as possible as soon as possible. In both instances the government's action is likely to be significantly determined by economic considerations.

> Although economic policy is only one aspect of Government policy it is obvious that unless it is sound it will push all other policy considerations into the background. The achievement of balance in political, social and economic objectives, setting the demands of economic growth into context with demands of defence, social justice and political acceptability, may prove quite impossible if a government's economic policy fails.
>
> (Preston, 1967)

There is a limit to which governments can afford to ignore economic principles; this limit is extended when strategic considerations are important, but it is clear that a government cannot afford to subsidise a region or an industry on a major scale for very long, or the process may become too expensive and have repercussions on the whole state. It must be stressed that although governments will give special consideration to areas, sectors and sections of the state, they will do so only within the context of the national situation. The intention is never to make those parts of the state self-contained; it is to enable them to make the maximum contribution to the well-being of the entire state.

Such policies will seek to modify the effects of geographical factors, such as distance and resource quality, and economic factors such as efficiency and capital availability, and they may be designed to operate at any stage of the production process. For example, if production, whether in agriculture or manufacturing, is considered to consist of the three stages, assembly of materials, processing and marketing, government policies can be found which will benefit different producers at different stages. Farmers are assisted in the assembly of materials in certain countries through the provision of subsidised fertilisers and the construction of major irrigation works; similar assistance is given to manufacturers through the operation of reduced tariffs on necessary imports, and tax rebates on funds expended in training skilled labour. In the processing stage, farmers may be helped by an advisory service, while the manufacturer may be helped by the provision of cheap power. The marketing of both manufactured and primary products may be assisted by tariffs which exclude competitors, by subsidies which assist competition in export markets and by quotas which restrict similar imports. In arranging the subsequent material four methods were considered: arrangement by economic activity, such as agriculture, mining and manufacturing, arrangement by stage of production such as processing or marketing, arrangement by the geographical factors which were being modified, such as distance, and arrangement by the policy concerned. All have some merit and all involve some measure of duplication, but organisation by policy was selected on the grounds that this would involve the least duplication and would focus on the policies to which geographers should give more attention. There is only one reservation which must be noted about this arrangement. Government policies are rarely isolated acts; in many cases the government will attack a particular problem from a number of directions, and will ensure that each policy plays a co-operative role with all the others. For example, if a government decides to maintain the price for a particular farm product, it may do so in any of three ways, and may choose to employ all three at the same time. Production can be restricted to create an artificial shortage; prices can be subsidised when they fall below a certain point; demand can be stimulated. This means that while policies have been separated here to show their effects more clearly the student studying a particular situation will have to examine any interaction of policies which may occur.

It is proposed to consider specific unilateral policies in four main groups. First the *regulations* established by governments will be considered; second, government *investment* will be examined; third, *financial* policies will be studied; and fourth, reference will be made to *regional plans.*

Regulations established by government are direct prescriptions which may determine where a particular activity occurs, the volume of production or import which is permitted, and how a particular development may occur. Such regulations differ from financial policies where the government hopes that a particular inducement will produce a desired effect.

Many governments have created regulations which determine where particular activities may be located.

What the Government's policy is directed towards is ensuring that there is a higher level of the use of resources in Wales, Scotland, the Northwest and the North, without overheating in Lewisham, London, the Southeast and the rest.

(Mr Callaghan, Chancellor of the Exchequer, London *Times*, 8 November 1967)

. . . the Minister of Planning may now refuse to allow any expansion of industry in existing industrial complexes like the Rand. He can do this by the stroke of a pen. By freezing economic growth in established White areas it is hoped to boost the border areas still further. For the choice that industrialists could face is either to mark time where they are, or expand in border areas.

(*Financial Mail*, 14 July 1967, p. 61)

In Britain the Board of Trade must approve all projects to build a factory exceeding 5,000 square feet in area, which means that the government can ensure that the location conforms to its plans for the future development of industry. This policy is backed up by the fact that grants may be made to the capital cost of establishing the factory in areas which have a high, or potentially high, rate of unemployment, classified by the Local Employment Act of 1960 as being those areas where the rate of unemployment has increased by 4½ per cent (Bowden, 1965, p. 30). Goodwin (1962) has studied the operation of this policy on the location of the motor industry. He describes how new development is being concentrated on Merseyside in England, at Cardiff and Llanelly in Wales, and in Linwood and Bathgate in Scotland. He concludes that the scale of government assistance given will ensure that if these sites are found to be completely uneconomic the firms will escape with little loss.

In South Africa, control over industrial location is being increasingly identified with the government's Border Area Development policy. This policy is designed to create work for Bantu near their Homelands, so that the drift to the towns in White areas can be halted and in time reversed. Since 1960 the government's policy has been increasingly applied by the Industrial Development Corporation, which was established in 1940 to promote industrial development. It was natural that the Corporation should be given this task, because it has long been an advocate of decentralisation. This policy has been compounded of many parts. First, as the quotation above suggests, it is being made more difficult to establish factories in established White areas. Proposals for such projects must satisfy the Departments of

Commerce and Industries and Bantu Administration, and the Permanent Committee for the Location of Industry that it is absolutely necessary, that there is a satisfactory site, and that there is a favourable Bantu-White ratio of workers and that the Bantu workers could be recruited from existing supplies in the area. Suitable sites are likely to become increasingly scarce through the operations of the Physical Planning and Utilisation of Resources Bill (*Financial Mail*, 14 July 1967). Second, there are the positive inducements to establish industries in the border areas. For example, the cost of water to industries in border areas will not be based on the cost of providing that water through the construction of headwater installations; low interest loans will be given to local authorities wishing to construct shunting yards and railway spurs; power, water and transport costs will be subsidised at various rates; a proportion of capital investment will rank as tax deductions in most cases; the Industrial Development Corporation will build factories to exact specifications and rent them to the manufacturer; and allowances will be given for the training of Bantu. By mid-1967 new industrial areas were being established at Hammarsdale in Natal, Rosslyn, near Pretoria, Pietermaritzburg, Ladysmith, Rustenburg, Pietersburg, Tzaneen and East London. By that time R180 million had been invested in the border area projects, providing work for 52,000 people, of whom 78 per cent are Africans.

Linge (1967) and McKnight (1967) have both made useful contributions to the study of the role of government in industrial location in Australia and South Australia respectively. Linge shows the importance of considering relationships between particular state governments and particular major firms, such as Broken Hill Proprietary Limited and General Motors Holden Limited. He believes that competition between state governments for major industrial projects has adversely affected the decentralisation policies of each state.

Regulations concerning the volume of production may be conveniently described as quota policies. For example, in Australia margarine manufacturers are only allowed to produce and sell a certain amount of table margarine; each firm has a fixed quota and penalties may be enforced if the quota is exceeded. This imposition is intended to avoid excessive competition with butter produced by dairy farmers, who in any case receive a direct subsidy and enjoy a price equalisation scheme so that they may export to Britain and Europe. It is important to look beyond the industry immediately affected by the regulations, to others which may be indirectly affected. In the case of the margarine quotas in Australia, it is argued by the manufacturers, with some justice, that the policy in effect places a production quota on the safflower oil industry, which will reduce the acreage planted from 100,000 acres in 1966–7, when one major margarine producer ignored quota restrictions, to 35,000 in 1967–8, when the restrictions were being enforced.

Quotas are often established on imports either by quantity or value. Towle (1956, chapter 24) has made an illuminating analysis of the development of import quotas. His study shows that quotas on various commodities are introduced for a variety of reasons. For example, quotas on agricultural products are largely designed to reduce imports, protect domestic prices and preserve the living standards of the peasantry. Industrial goods may be subject to quotas to preserve the domestic markets of industries which are vulnerable to foreign competition. Luxury goods will be subject to quotas when attempts are being made to improve the balance of trade. Grotewold and Sublett (1967) have shown that import restrictions may be one of the factors which influence changes in the intensity of landuse in Britain and Germany, and there are sound reasons for thinking that their conclusions may have a more general application.

Quotas have also been established for agricultural output from time to time in various countries. The quotas may be either acreage controls or marketing controls. Under the acreage quota each farmer is awarded a certain acreage, which may be planted to the crop in question. In America, farmers who co-operate in such schemes, in respect of cotton, wheat, hogs and tobacco, receive benefit agreements or price guarantees (Schickele, 1954, pp. 199–200). Under the price marketing scheme the farmer is only allowed to sell a certain amount of a particular crop or product. This is a much more precise scheme since variable yields do not influence the situation. In America there are penalties for farmers who sell more than the quota; these penalties are set so that they reduce the profitability of such surplus output. Agricultural quotas are fairly common as a recent survey of the United States Department of Agriculture (1964) shows. For example, Spain has acreage quotas for wheat, rice, and sugar, both beet and cane; Sweden has acreage quotas on sugar beet and fibre crops; Lebanon has acreage control on tobacco, while Rhodesia controls tobacco through a marketing quota; and Greece controls tobacco and currant acreages.

Lastly, governments may impose certain quotas on the proportion of imported materials which can be used in the manufacture of a particular product. This has been the policy of the Australian and South African Governments to stimulate their domestic motor industries. In South Africa the manufacturer must employ at least 55 per cent of local materials in the production of the car by value. These regulations became effective at the beginning of 1968. In Australia the local content proportion is higher and this has been responsible for the decision of a number of overseas firms to establish plants in Australia, and for the thriving motor-parts industry in Australia, which is larger than might be expected in view of the number of vehicles in the country.

It is more difficult to discover examples of a government regulating the way in which a resource should be developed, but instances have occurred in the various rules governing oil exploration throughout the world. Australia is one of the latest countries to discover major sources of petroleum and natural gas, and in 1967 legislation was introduced into the federal and six state parliaments to control the exploration and production of oil from off-shore areas. One section of this legislation refers to a quadrant concept, which requires any firm which discovers oil or natural gas to yield four-ninths of the lease to other firms. This policy is presumably aimed at promoting rapid development of such resources. Much of the significance of the section has been removed, however, by the inclusion of a clause which allows the original discoverer to keep the entire lease providing a higher royalty is paid on production from areas which would have otherwise passed to other firms.

This survey has not covered all possible government regulations, but it has considered the most important from the point of view of the political and economic geographer.

The second major group of specific unilateral policies relates to government investment. The significance to geographers of government investment is that the government will decide *what* the money should be spent on if there are a variety of conflicting projects, and *where* the investment will occur if there are a number of possible sites. It is not proposed to consider such policies in detail because geographers have shown a real appreciation of the significance of government investment. In *Geography* the section entitled 'The Changing World' regularly contains factual notes showing the significance of a new dam, or railway line or port. Longer papers have also appeared at regular intervals dealing with projects such as the Kariba Dam (Reeve, 1960), and irrigation and transport developments in the Riverina of New South Wales (Rutherford, 1963; and Smith, 1964). A significant number of papers have been published dealing with the attempts of governments to transform subsistence agriculture; the most important include studies of Swaziland (Daniel, 1966), Zambia (Richards, 1958), Matabeleland (Prescott, 1961) and Nigeria (Coppock, 1966). One of the most useful papers to appear recently has dealt with irrigation and internal colonisation in Spain. Naylon (1967) stresses the importance of basing any understanding of the Spanish landscape on the national legislation and government agencies of the country. In a carefully documented account he shows, so far as he is able, the reasons for various government regulations and decisions concerning investment, and the consequences which followed these decisions in the transformation of the agricultural landscape, distribution of population and farm production. This paper could well serve as a model for other geographers interested in similar subjects.

The third group of specific unilateral policies relates to financial measures, which generally are formulated to encourage development by private enterprise along certain lines considered desirable by the government. The policies which will be considered here are differential exchange rates, subsidies, taxes and tariffs. For an economic discussion of the theory of such measures geographers must turn to books written by economists and political scientists. The most useful of these studies include those by Culbertson (1924), Towle (1956) and Woytinsky (1955); recently Thoman and Conkling (1967), two geographers, have produced a study of international trade which includes a helpful section on tariffs and a technical glossary.

The manipulation of exchange rates to favour a particular area or a particular industry is much more a characteristic of under-developed countries than developed states. Baer (1964) has discussed the Agio system in Brazil which requires importers in the north-east part of the country to pay higher rates for foreign exchange than those in the southern region.

> Proceeds of these rates have been used by exchange rate authorities to prop up the coffee economy, which is centred mainly in the south. Excess balances from the Agio system also increased the capacity of the Banco do Brazil to make loans, a high proportion of which are made in the south.
> (Baer, 1964, p. 281)

Moes (1966) has examined a similar situation in Central Africa, when Rwanda and Burundi formed a customs union. The difference in exchange rates in this case was not between two areas, but between two classes of goods. Importers purchasing goods classified as necessary were allowed to obtain foreign exchange at the official rate which was fifty francs to one American dollar; foreign exchange for other goods had to be obtained on the open market where the rate was one hundred francs. Exporters had to surrender their foreign exchange earnings at the official rate, which amounted to an export tax. Moes has shown that these arrangements affected different economic enterprises and different racial groups in different ways, and was a major underlying cause for some of the political problems which developed between the two countries. Leftwich (1966) has discussed the operation of multiple exchange rate systems in Chile from 1940 to 1959, although he does not show the way in which this might have acted differentially throughout the country.

Subsidies are intended to increase the profitability of an operation, either by direct payments or by reduced prices. Some economists would argue that a protective tariff was also a subsidy, but such arrangements will be considered in the next section. The distinction given here means that the subsidies considered refer mainly to the farming industry, whereas the subsidies to manufacturers, made available through tariffs, are considered later.

The subsidies made available to farmers through lower prices for phosphates and preferential freight rates are well known; price support programmes have received less attention from geographers, although they are no less significant in helping to explain why certain patterns of production exist in certain countries. Schickele (1954) has made a detailed study of price support programmes. Those involving subsidies include surplus production purchase, export subsidies and price deficiency payments. Surplus production purchases mean that the government will buy any stocks which cannot be absorbed into the market, but such policies require control measures along the lines considered earlier. Export subsidies involve a discrimination between higher domestic and lower overseas prices. For example, in the case of Australia, this means is the most frequent form of price support (Lewis, 1967). It operates, in connection with other schemes, for wheat, dairy products, dried vine fruits, sugar, rice and eggs; in the case of milk, sugar, eggs and peanuts discrimination also varies between the use made of the product. The equalised price which is paid to the farmer is the average net revenue from all markets. Lewis (1967) has noted some of the disadvantages of this scheme.

> As production responds to higher prices erosion of benefits of multiple-price schemes occurs, unless the home market grows as fast as production and export returns are not depressed by the additional quantities sold. Moreover the differences between the equalised price and the industry's marginal return tends to be capitalised into fixed asset values, especially land and livestock. The result is that alternative enterprises, such as forestry or beef production in dairy areas, often face inflated factor prices, which impede land-use adjustments and distort interregional competition in these products. Incentives for marketing improvements and for aggressive competition also tend to be reduced.
>
> (pp. 306–7)

Governments also make direct payments to farmers. For example, in Australia the government pays an annual subsidy of $A27 million divided on a proportional basis amongst all dairy farmers whose products enter the equalisation scheme. There is also a bounty on cotton which varies with quality; again a ceiling has been fixed on the bounty, in this case it amounts to $A4 million. Schickele (1954, pp. 213–15) describes the payments made to American farmers and he shows how the proportion of farm incomes paid as direct subsidies varies regionally throughout the country at different periods of the programme.

Taxation is an obvious means of influencing economic development and many governments use it. In South Africa large tax incentives have been offered to the first firms which find oil and natural gas in on-shore and off-shore areas; in Australia, calls on mineral shares for the purpose of oil exploration are tax deductions while dividends from gold shares are tax-free. A novel tax introduced by the British Government taxed industries which could be classified as service harder than those described as manufacturing. The announced intention of this policy was to encourage a transfer of labour from service to manufacturing industries.

Tariffs are imposed on imports for one of two reasons. In certain cases the tariff is designed to give protection to a domestic industry which might suffer from overseas competition; governments thus seek diversification and a measure of self-sufficiency through tariff schedules. In other cases governments impose tariffs on imports and export duties on exports in order to raise revenue. The apparent neglect of tariffs by economic geographers, judged by the major textbooks and recent articles, has been partly redressed by the attentions of Grotewold (1960) and Thoman and Conkling (1967). Thoman and Conkling devote a whole chapter to the logistics and mechanics of international trade and provide a useful bibliography showing where further, more detailed information is located. Geographers will study tariffs for different reasons, but those interested in policy decisions will seek to establish the reasons why a particular tariff was established, why the tariff was fixed at a particular level, and the consequences of such a tariff on the country which imposes the tariff and the states whose exporters try to hurdle the tariff obstacle. The two basic reasons have been outlined above but there are two others which should be mentioned, even though they are less important. First, some governments are undoubtedly concerned about the conservation of natural resources and believe that it would be unwise to allow the national reserves of strategic minerals to be exhausted. In some cases the refusal to impose tariffs requested by domestic producers is based on this reasoning. In other cases governments prefer to establish stock-piles at a level which would ensure against any foreseeable emergency. Second, tariffs can be a political weapon; they may be imposed so that they can later be removed as a concession, or they may be imposed to record disapproval of the actions of some particular government.

One of the most useful studies of commercial policies of states was prepared by Culbertson (1924). In his analysis of tariff policies he recommended that three main groups of commodities should be studied. First there are those items on which the state is wholly dependent on foreign supplies; second there are those items where the state produces enough to satisfy its own requirements and in some cases will have a surplus for export; third there are the items where the state's domestic supplies have to be augmented by imports.

In the case of the first group it follows that the consumers who need the item which must be imported are anxious to obtain it as cheaply as possible. Under such cases governments will not impose tariffs unless they wish to raise revenue, or, in the case of a luxury product, curtail the level of imports, or, in the case of a mineral, encourage exploration to discover domestic supplies.

The last two aims could be more easily achieved by a prohibition or quota against luxury goods, and by means of incentive payments, such as tax rebates in the case of mineral exploration.

The second group does not permit any general prediction and the position will vary with individual circumstances. For example, if the item produced domestically is more costly than the imported item then a tariff may be necessary to protect the domestic producer. On the other hand, if domestic production is very competitive in price and there is a surplus for export then there will be no need for a tariff on protection grounds.

The third group of items, where domestic production must be supplemented by imports, also provides a complex pattern. Culbertson (1924, p. 11) introduces the helpful concept of reproducible and non-reproducible items. Reproducible items are those whose production can be expanded; Culbertson was particularly concerned with raw materials, but this would be equally applicable to manufactured items. Thus production of crops such as cotton and wheat could be expanded, assuming land in a suitable climatic belt is available, in a fairly short space of time, whereas timber production could not be quickly increased, unless there were large unused forests. But if a country has limited coal reserves then it cannot expand production to eliminate the need for imports. The government must also watch the conflicting interests of both producers and consumers. In the case of non-reproducible items then the government is unlikely to impose a tariff which would only operate against the consumer, although if the domestic producer provides a more expensive product than the imported item, it may be necessary to give some protection to prevent imports from acquiring the whole market. In the case of items which are reproducible governments may impose a tariff to promote expansion of domestic capacity and the gradual elimination of imports.

Culbertson maintains that the most important point revealed by his discussion is that

> import tariffs are determined not according to some general theory of free trade or protection, but by the conditions of each particular case in accordance with the national need.
>
> (Culbertson, 1924, p. 21)

This is a point which geographers must remember and which makes the subject a suitable one for their study.

Export duties are not considered by Thoman and Conkling (1967), but they are worth mentioning since they serve precisely the same function as a tariff for raising revenue. Export duties tend to be levied by states on raw materials rather than on finished products, which means that they are more commonly encountered in the case of under-developed countries. Towle (1956, pp. 439–45) explains that such countries often find it easier to collect revenue from such a source, than from personal income tax; during the 1930s Chile obtained 80 per cent of its revenue from export duties. Normally duties will be concentrated on the most important items; in India tea and jute are subject to export duty, in Chile sodium nitrate, in South Africa diamonds. In some cases this duty may be imposed in order to make the raw material so expensive that the manufacturer will decide to establish a factory within the country. Spain and Portugal are generally considered to have levied duties on the export of raw cork in order to promote the development of a local cork industry. The United States Department of Agriculture, in a survey of agricultural policies in foreign countries (1964), made frequent reference to export duties in the case of countries such as the United Arab Republic, Ghana, Kenya, Mozambique, Malagasy and Malaya.

Geographers are interested in the variation of tariff levels to meet particular situations. For example, import tariffs may be reduced if the whole country is in a condition of need, or if a particular industry is in difficulties, or if a particular area faces serious problems. In February 1957 Ceylon reduced the import duties on coconut meal to overcome shortages and rising market prices; in July 1957 Chile suspended duty on the import of edible oils for a brief period for the same reason. In 1957 the building industries of Syria, Cuba and Portugal were helped by reductions in the import duties on glass and cement; the mining industries of Italy, Chile and Peru were helped by reductions in the tariffs on certain fuels and machinery; and manufacturing industries were helped in Brazil (motor vehicles), Dominica (glass) and Nigeria (all new industries) by reductions in the tariffs on the import of capital equipment. Finally, Argentina abolished import duties on all items required for the industrial development of Patagonia, while France took the same action for the benefit of the Saharan territories of Algeria (General Agreement on Tariffs and Trade, 1958, pp. 9, 11 and 12).

Import duties may also be raised when there is any danger that dumping by other states may occur. Dumping refers to allegedly unfair trading. For example, the Customs Tariff (Dumping and Subsidies) Act, 1962 passed by the Government of Australia defines dumping in the following terms:

> . . . dumping duty may be imposed on goods that are sold to Australian importers at a price which is less than the normal value of the goods, where this causes or threatens material injury to an Australian industry. 'Normal value' under the Act means:
>
> (a) fair market value in the country of export;
> (b) price in the country of export to a third country;
> (c) fair market value in a third country; or
> (d) cost of production, plus f.o.b. charges, plus selling cost and profit.
>
> Countervailing duty may be levied on goods in respect of which any subsidy, bounty, reduction, or remission of freight . . . has been paid.
> (*Commonwealth Bureau of Census and Statistics*, 1966, p. 399)

Sometimes the anti-dumping legislation refers to a specific item imported from a nominated country. For example, in 1957, South Africa enacted anti-dumping tariffs in respect of acetone from the United Kingdom, fabrics and woven piece goods from the United States, electrical motors from West Germany and cotton piece goods from Japan, Czechoslovakia and Hungary. Belgium in the same year raised tariffs on pure and artificial silk imported from Hungary or Poland. In Australia and Canada the governments have passed general legislation to cover anti-dumping activities without specifying the source.

Preferential tariffs have not been considered here because they are multilateral policies.

The remaining type of specific unilateral policy concerns regional plans which governments design from time to time. The policies which together constitute a regional plan, such as those dealing with the control of industrial location, differential taxation and exchange rates, have already been considered. But the importance of regional plans is that they are concieved as a whole, and because of the interaction of one policy on another they must be considered within the context of the total plan.

Regional planning is apparently best developed in totalitarian states but examples can also be drawn, on an increasing scale, from the democratic states. *Soviet Geography* has published many papers dealing with the territorial organisation of economic development; recent papers by Gerasimov (1964), Saushkin

(1964) and Tokarev (1962) are amongst the most interesting. It is clear that in Russia there is a tendency to regard the internal boundaries of *oblasts* and other areas as being less rigid; planning boundaries are breaking away from this regimentation and producing new, more realistic patterns. The paper by Gerasimov and others displayed a critical spirit which listed some of the defects of earlier systems, including the stereotyped application of processes developed in other areas, the inability of some planners to see the consequences of their policies, and the excessive concern with short-term aims at the expense of long-term requirements.

There is a wide literature dealing with regional planning in Western countries; much of it is considered in the books by Freeman (1967), Jackson (1966) and Minshull (1967). Government reports, such as the White Papers issued by the British Government dealing with north-east and south-east England, Central Scotland and Northern Ireland, provide useful starting points for geographers studying regional plans; they also raise the question, referred to in the first chapter, about the extent to which geographers should become involved in policy-making. Thomas has no doubts about this question.

> The problem of distressed areas represents a complex of economic, social and political relationships which can be properly understood and assaulted by co-ordinated research efforts and planning by members of many disciplines. Because of their regional perspective and their training in area analysis, geographers are clearly well qualified to participate in attempts to resolve redevelopment area problems through research or in advisory capacities. An opportunity is at hand for geographers to distinguish themselves in the work of regional planning. . . .
>
> (Thomas, 1962, p. 15)

Thomas goes on to stress that geographers must be aware that when they give advice it may easily be directly translated into policy and therefore they must be willing to accept full responsibility for this situation. Gerasimov (1964) also makes a plea for geographers to be consulted by planning and policy-making organisations. The special contributions which geographers should be able to make are four in number. First, geographers are better equipped than any other worker to define regional boundaries, whether the search is for distressed areas or the best territorial units for economic development. Second, geographers have the ability to describe the existing situation within such regions—the base on which the plan will be superimposed. Third, geographers are capable of showing how previous plans have produced certain predicted and unpredictable effects on the landscape, and of using this experience to outline how new policies may react upon the landscape. Fourth, geographers have a capacity to see the region as part of the larger state; this helps them to explore the interrelations between the regional and overall national plans.

General multilateral policies

These policies are determined by at least two states for the benefit of the entire economy of at least one of them. It was noted in the third chapter that states often seek to increase state strength through alliances. This is even more true in the sphere of development, because while there are some states which have no formal defence treaties with other states, all states have at least some formal commercial agreements with other governments. This is understandable since no state which aspires to develop economically can rely entirely on its own resources; imports must be obtained and they must be paid for through exports of one kind or another. The definition which opens this section refers to benefit to the economy of at least one of the states concerned. This may suggest a measure of altruism in international commercial policy but this was not intended. Between states of equivalent wealth commercial agreements will normally benefit the economies of both states in some measure. It would be clearly possible for a state, which was anxious to obtain some favour, such as transit rights, to offer very favourable terms to the state which conferred these rights, so there is no need to assume that the benefits must be exactly equal. On the other hand, where there is a major discrepancy between the economic strength of two states it is quite possible that the commercial arrangements will appear to offer very little advantage to the stronger state; this is the case in many aid programmes established by the wealthy states. But in such cases the wealthier state may be obtaining certain strategic or political benefits rather than economic advantages.

This discussion raises one interesting point. It is often assumed that useful commercial agreements can only be concluded between developed industrial states, or between developed and under-developed states. Co-operation between under-developed states is often dismissed because of their competitive rather than complementary forms of production. There are, however, a number of ways in which under-developed states can help each other despite the similarities of their problems and the competitive nature of their exports. Prescott (1966) reviewed some of the policies between under-developed states in West and North Central Africa. They included customs unions, transit rights for land-locked states, bulk ordering of imports by a number of countries, co-operation regarding the investment of aid, and the joint development of trans-boundary resources.

Four types of commercial association will be considered in this section: international organisations, customs unions, preferential tariffs and general aid programmes.

Amongst recent general texts in political geography, Pounds (1963) and de Blij (1967) provide useful sections dealing with international and regional economic associations. The main international economic agreements are associated with the United Nations: the International Monetary Fund, the International Bank for Reconstruction and Development, the International Finance Corporation and the General Agreement on Tariffs and Trade. The International Monetary Fund had 105 members in December 1967 and its aims were to promote international monetary co-operation and exchange stability, to assist in the removal of exchange restrictions, and to help the expansion of world trade as a means of encouraging high levels of employment income and resource use. The International Bank for Reconstruction and Development had 105 members in December 1967 and its aim was to promote and facilitate international investment to increase production, raise living standards and improve the conditions of world trade. The International Finance Corporation had eighty-two members in December 1967 and its aim was to stimulate private investment in under-developed countries, without government guarantees. The General Agreement of Tariffs and Trade had seventy full members in December 1967 and it aims to promote improved standards of living, full employment and economic development through fiscal measures relating to world trade. It is apparent that these four agreements have similar general aims but that they concentrate on different aspects of the problem of world economic development. Further it is apparent from the published tables of the loans made, and the tariffs lowered, that while they may not have achieved all their sponsors had hoped, these agreements are powerful forces for economic change throughout the world. For this reason alone it is clear that geographers should study them. Economic and political geographers share this interest; but while the economic geographer will be most concerned with these agreements as factors which alter production and trade patterns, political geographers must also be concerned to explore the geographical reasons which prompt a state to avail itself of the advantages of membership or to withhold its support.

In addition to these international agreements there are also certain regional associations which supplement them. The Colombo Plan was established in 1951, and in December 1967 had twenty-four members. Most of the states represented are contained in the area between Japan and Iran, but there are extra-regional members including Britain, Australia, New Zealand and the United States. This voluntary organisation aims at the economic development of the under-developed members through co-operation which includes aid programmes from the wealthier states.

Turning from these large international and regional organisations it is necessary to consider the smaller regional associations which usually involve a much closer form of economic association. These associations have a variety of relationships, but normally agreements will be reached which will reduce the barriers to circulation of goods, capital and people amongst the members. The European Economic Community, the European Free Trade Association, the Customs Union between South Africa, South West Africa, Botswana, Lesotho and Swaziland, the Equatorial Customs Union of Central African Republics, Congo, Gabon, Chad and Cameroun, the Entente formed by Ivory Coast, Dahomey, Upper Volta, Niger and Togoland, and the Latin American Free Trade Area are all organisations of this type. Pounds (1963, p. 302) has shown that these organisations evolve slowly; barriers to circulation are not removed abruptly, but at a rate which will allow the units of production, factories and farms, within each state to adjust to the new circumstances. This gradual development affords an opportunity for the geographer to measure and identify the changes at various points in time, thus gaining a clearer insight into the processes involved. It is important that this should be done since in some cases statistics relating to intercourse between member states of a single customs union will cease to be available, making comparison with earlier conditions difficult.

The levels at which tariffs are set often varies with the country concerned: they are fixed through bilateral negotiations between the two countries which seek to gain the maximum advantages for themselves. It follows that if a state can arrange for lower rates of import duties on products which it is trying to export, its producers are given an advantage compared with the producers of other countries who have to pay the higher rates. Usually such arrangements must be reciprocal, but there have been cases, after a war, when the vanquished state was required to afford certain tariff benefits which were not provided by the victorious state. This was the case in the 1856 Treaty between Siam and America and the Treaty of Versailles between Germany and the Allies. States which negotiate reduced tariffs will often try to include a clause which ensures that no other country is treated more favourably. This is known as the most-favoured-nation clause, although it has been pointed out by a number of authors that the purpose of the clause is not to establish a more favourable position than any other state, as the name suggests, but rather to guarantee equality with the state which is most favourably treated. In many cases this condition will apply to all categories of goods from all other states, in other cases it will apply to only a single item or group of items, and sometimes it will apply only to other agreements with nominated states. For example, in the Franco-German Peace of Frankfort in 1871, the most-favoured-nation clause only applied to agreements reached between either country and the following states: England, Belgium, Holland, Switzerland, Russia and Austria. This most-favoured-nation clause is interesting, in that a state may acquire benefits at no cost, which were secured by another state possibly at significant cost. In many cases the most-favoured-nation clause is not deemed to apply to traffic of two kinds. First, there is the case of traffic within a borderland. Many states, recognising the problems of living in a borderland divided by an international boundary, allow free traffic across the boundary by recognised border dwellers. This advantage is not extended to traffic from outside the nominated borderland, and it is not considered to apply under the terms of a most-favoured-nation clause. Second, tariff concessions between states which enjoy close and special political or ethnic relationships are usually excepted from the operation of the most-favoured-nation clause. This is true in the case of relations between imperial and colonial areas. In many cases the preferential tariffs established between such areas have been continued in some form after the colony has achieved independence. The simple system which applied in the British Empire until the middle of the nineteenth century involved free entry or reduced tariffs on raw materials entering Britain, and similar arrangements for British-manufactured articles exported to the colonies. This pattern has become much more complex as colonies have become independent, as some have developed closer trading contacts with non-Commonwealth countries, and as some have developed a significant sector of manufacturing industry. Any geographers dealing with international trade by itself, or as an aspect of the political relations between states, cannot escape from the need to explore the various tariff schedules as they apply to various states, and as they have altered from time to time.

The last policy considered in this section concerns the provision of aid from states which can afford it to states which need it and are prepared to accept it. There is a very real distinction between aid which is provided for the general development of a country and aid which is provided for a particular purpose. Most foreign aid is of the second type and is considered under the heading of specific multilateral policies; it therefore seems appropriate to deal with the geographer's interest in this field in that section. It is sufficient at this point to draw attention to the problems of measuring the volume of aid. It can be given in many forms: through direct cash grants, through loans under various conditions of repayment, through technical assistance, through the training of students, and through the sale of food at uneconomic prices. It is rarely possible for the geographer to measure these diverse forms of aid in a common unit, yet there must be an awareness of each, since they may all have an important influence on the economic development of the receiving country. This problem may have accounted for the slight consideration which aid programmes have received from geographers; Black (1963), Scheel (1965) and Symons (1967) may be regarded as pioneers in this field. Black reviewed the pattern of American aid to African countries and concentrated on identifying the six main guide-lines of American policy. These were that the long-term aid was economic development which would require comprehensive planning by the African states, that aid should be concentrated in those areas where there was likely to be the most rapid development, that loans should be made in preference to grants and that financing should be through dollars for the purchase of American goods and services, that American aid should be co-ordinated with the aid of other countries through international agencies, and that American aid should normally be subsidiary and subordinate to aid from the former colonial power in the case of the states which recently achieved independence. Scheel (1963) surveyed the contribution which West Germany has made to African states, which has been largely devoted to developing the economic and technical infrastructure needed for more rapid development, and the establishment of industries which can process primary products originating in African states. He makes the important point, which is applicable to most financial policies, that they cannot alone guarantee favourable results, they can only create a favourable situation for economic advance, which will rest on the attitude which the population adopts. Symons (1967) examined the provision of Russian aid to the under-developed countries of Asia and Africa. He was able to show that Russian aid is much more concentrated in its application and is more frequently used in the establishment of major capital works such as the Aswan High Dam.

Specific multilateral policies

These policies, which are determined by at least two governments apply to a particular area of the state, or sector of the economy, or section of the population. There are three kinds of specific multilateral policy which are of particular interest to geographers: trade agreements in respect of particular items, the provision of aid for a particular purpose, and the joint development by two states of a common resource.

In considering specific trade agreements it is important to draw a distinction between bilateral agreements dealing with a single item and international commodity agreements. Bilateral agreements between states in respect of particular items usually relate to the tariffs which will be imposed, and normally there will be mutual advantages. Further, the item can be a manufactured commodity or a raw material. For example, in June 1957, Cuba reduced import tariffs on tinplate, tin sheets, artificial colours and motors from the United States, and in return the United States reduced import duties on five kinds of cigar tobaccos (General Agreements on Tariffs and Trade, 1958, p. 6).

International commodity agreements are concluded by a number of states, both producers and consumers, to reduce the severe fluctuations in the prices of primary products which occur from time to time. Rising prices, consequent on excess demand, create problems for consuming nations, while falling prices, consequent on excess supply, create difficulties for the countries which produce the raw materials. On balance the latter situation is probably the most harmful in international terms for three reasons. First, the price rise can usually be more easily borne by the developed states, which are the major consumers of raw materials. Second, because in times of short supply it is possible to use substitutes in some cases; for example, a significant rise in the price of copper encourages the use of plastics and aluminium. Third, the under-developed states, which are major suppliers of many raw materials, have vulnerable economies which are ill-equipped to cope with a severe fall in the price of staple exports; in such cases economic problems may have a real influence on the political and social structure of the state. While commodity agreements are designed to remove a common problem there are a number of factors which make it necessary for geographers to look at individual cases with care. The importance of commodity agreements will vary directly with the amount of the raw material which enters world trade. For example, during the period 1955–9 the bulk of tin (98 per cent), rubber (96 per cent), cocoa (88 per cent) and coffee (77 per cent) were exported from the sources of production; by comparison only 16 per cent of the world's wheat production was exported during the same period (Hudson, 1961, p. 507). The number of major producers involved varies from commodity to commodity and clearly affects the ease with which agreement might be reached, although the number of states is not the only factor to be considered in this case; the *attitude* of major producers is at least as important. Lastly, the significance of these commodities to the economies of individual countries varies, giving each different degrees of vulnerability.

> Out of more than thirty years' study of commodity problems and the subsequent application of international commodity agreements, three distinct types of agreements have emerged. These are the 'export quota type', adopted as the basis of the Sugar Agreement, the 'buffer stock' type used in the Tin Agreement, and the 'multilateral contract' type which is employed in the Wheat Agreement.
>
> (Hudson, 1961, p. 510)

The export quota system seeks to restrict the amount of the commodity entering world trade, in order to match demand as closely as possible; such a system can only operate if all the major producers subscribe to it. The 1958 Sugar Agreement specified the basic export levels of each exporting member, defined how alteration of quotas could occur if there was a rise in price, or if certain states could not fulfil their quota, and stipulated the amounts of sugar which importing countries were allowed to produce or import from states which were not party to the Agreement. The buffer stock system sets a floor and ceiling price. The authority must purchase the commodity when the price falls to floor level and sell when it reaches the ceiling. This method faces the danger that stocks will mount if the price remains near floor levels for a protracted period; export quotas may be introduced to prevent this happening. Multilateral contracts also fix a price range, but in this case the importing countries guarantee to purchase a fixed proportion of their needs at prices above the minimum level, while exporting countries guarantee to supply a fixed quantity at levels below the maximum price.

Commodity agreements do not seem to have attracted much attention from geographers; the interesting paper by Courtenay (1961) on the impact of price agreements on tin production in Malaya is exceptional. Geographers interested in policy might profitably study such agreements to discover why states decide to co-operate or stand aloof, and what are the consequences of such co-operation or disagreement on the patterns of world trade and production. Scholars in other fields have studied these agreements in detail. Ojala (1967) has prepared an interesting survey of these agreements since the Second World War. He concludes:

> in terms of world commodity problems, the results of post-war efforts, useful though they may have been in many instances, can only be described as meagre.
>
> (Ojala, 1967, p. 36)

He identifies the major causes of this situation as being the national interests of states and the welter of bilateral trading arrangements. A fuller discussion is provided by Pincus (1967), an economist, who looks at the question in a geographical fashion, by considering trends and developments by commodities and countries. At the end of the survey he reaches four main conclusions, although he warns against any over-generalisation in view of the uniqueness of each commodity situation. These conclusions are:

(1) Many Latin American and Asian countries will be able to rely less and less on commodity trade for financing import growth.

(2) The Middle East and Africa are better off, thanks largely to petroleum in the former region and favourable supply conditions for a number of products in the latter.

(3) Two conditions will tend to favour the growth of any country's commodity export trade:
Low-cost productive potential, even if the particular commodity faces sluggish world demand (as in the post-war expansion of African tea and sisal exports).
Specialisation in products with good demand prospects (as in Peruvian fishmeal trade, African copper exports, Middle East oil, or Malayan tin).

(4) For most countries this last condition does not apply. Those that are semi-industrial (India, Brazil, Argentina, Mexico, Taiwan) can hope to shift increasingly from commodities to manufactured exports, although this is no easy task. Those whose industry is not yet established have even fewer trade alternatives. For them, unless tourism or other service industries can be developed, the foreign exchange limitation is likely to loom as a major obstacle to desired rates of growth.

(Pincus, 1967, p. 258)

The distribution of foreign aid for specific purposes has not received very much attention from geographers, even though it must be one of the most powerful factors accounting for landscape change in some under-developed countries. Geographers have reported on the consequences of aid spending, which have included, for example, new industries, extended patterns of communication and fresh sources of power, but there has been no consistent effort to probe the question more deeply. It seems proper that political geographers should seek to identify any geographical factors which encourage states to give aid or accept it; which influence the pattern of aid-giving in terms of area and time; which help to explain the degree of competition or co-operation between aid-giving states and between aid-receiving states; and which influences the targets selected for development within receiving states. Analysis of the effects of aid application does not end with description of the changed landscape; it is also necessary to discover whether the changed conditions have any effects on patterns of trade throughout the world, and whether there is any connection between aid and the political relationships of states concerned.

Clark (1967) examined the provision of American aid to Africa He considered first the effects of the application of aid on the receiving countries, and then he turned to the more difficult problem of trying to understand why America gave aid to particular countries at particular times. It is impossible to be certain that there are causal relations between political geographical events and the volume and direction of aid-giving, but it is legitimate to draw attention to the coincidence of changes in the pattern of aid distribution with changes in other factors which may be presumed to be significant. Clark identified five major aims of American aid policy in Africa: to win political friends, to assist economic and political stability, to provide countries with armed forces so that they can resist subversion, to satisfy humanitarian motives and to further commercial interests. He then constructed graphs to show the variation in aid receipts for each country since the inception of the programme, and tested these graphs against a number of factors. These factors were mainly related to conditions within America and the receiving state and the relations between them. The most significant factors identified by this study were the internal economic and political conditions of America, and the relationship between America and the African state concerned. America's balance of payments problems and the attitude of Congress were clearly important in determining the volume of aid, the conditions which were attached to the aid, and the country which received aid. While there was not a perfect correlation between cordial relations and the receipt of a high volume of aid, or between unfriendly relations and the absence of aid, Clark showed many cases where a decline in the volume of aid followed worsening political relations. By contrast, volume of aid showed no consistent relationship with aid received from other sources, including the Soviet Union, or with the movement of private capital in Africa.

The problem of establishing the volume of aid has been discussed, but despite this problem, the subject does seem a legitimate field for the political geographer to study. Aid is given and accepted through political decisions which may be influenced by geographical factors; aid is a means of transferring wealth throughout the world and is responsible for some major landscape changes; finally, the giving and receiving of aid are an obvious indication of political relations between the states concerned.

Joint development of a particular resource will sometimes

involve two governments which are not neighbours, as in the case of Nigeria and Mauritania which agreed to jointly work the latter's phosphate deposits for the benefit of both states. Such an operation is quite distinct from the use of overseas capital in the development of the resource, an event which does not necessarily require the formal agreement of two governments. More frequently, joint development will be concerned with the common resource of two neighbouring states. The most common forms of transboundary resources are water and minerals, principally petroleum. Prescott (1966 and 1967) has considered co-operation between adjacent states for the development of water resources in West and North Central Africa, and in isolated cases in Europe and North America. The joint development of petroleum deposits has been undertaken by Saudi Arabia and Kuwait. Formerly parts of their territories were separated by a neutral zone. This was divided between the two states in 1966, but they continue to share the oil revenues from the former neutral area. Reference was made earlier to joint mining operations in the Moroccan-Algerian borderland.

Geographers would also be interested in the joint development of an integrated transport system to serve a common borderland; such a development would be distinct from transit rights which states may arrange for the benefit of the whole country.

National plans

The individual policies which have been considered in this chapter will often form part of a national plan for the development of the state, thus although a geographer may have to separate these various elements of the national plan to gain a clearer understanding of their operation, it is important that at some stage the plan should be viewed as a whole since the policies will tend to interact upon each other.

The national plans which are so common today amongst the under-developed countries of the world originated in the Soviet Union, and national economic plans are still a feature of Communist states. The importance of national plans is that they frequently result in the publication of a considerable amount of information which generally consists of three parts. First, there is often an economic survey of the country; the International Bank for Reconstruction and Development has been responsible for the publication of many economic surveys on which subsequent national plans have been based. Second, the government will publish the general aims of the plan and identify the main sectors in which activity will be concentrated. Lastly, targets will normally be set. This means that the geographer can identify the geographical factors outlined in the initial survey which were deemed to be important, although it will always be necessary to check such surveys and be aware of any geographical factors which were not mentioned, perhaps for political reasons. It also means that the geographer gains an indication of those sectors of the economy which the government considers need strengthening and which will bring the best return. Again, such decisions should be examined to identify any cases where they may be based on political rather than economic grounds. Such a situation could occur where a state was composed of a number of different racial or tribal groups. Lastly, the geographer is given a scale of intended progress against which actual progress can be measured. Differences between prediction and actual result can be examined to see whether geographical factors made any contribution. There is one defect of national plans. Usually they will refer only to capital expenditure within the country, analysing the amount to be spent on new roads, factory extensions and similar projects. Usually there will be no reference to external economic policies which will be designed to supplement the internal policies. Thus plans for the development of a domestic industry may not also explain that import tariffs will be raised against similar goods produced overseas.

It is surprising that despite these advantages geographers appear to have paid so little attention to national plans. Apart from consideration of the various Soviet plans in some general texts on the area, recent papers by Blouet (1965), Bettelheim (1963) and Farmer (1961) are the best examples of such studies.

Blouet (1965) reviews the results of Malta's first five-year development plan (1959–64) and shows how the desire to diversify the economy, because of the impending withdrawal of British forces, was achieved through the expansion of tourism and manufacturing. The main policies employed were government grants, various fiscal attractions to overseas capital, tariff protection and tax benefits. Bettelheim's paper on India's third five-year plan adopts the approach of the economist rather than that of the geographer. There is a clear statement of the targets, and the money spent, and the balance of payments problem, but at no stage is this related to the way in which different regions of India benefited in different ways, nor were there apparently any significant geographical facts which accounted for some of the problems which were becoming evident. Farmer's paper is the best balanced of the three. He shows the situation of Ceylon before the initiation of the plan, and lists its major objectives and assumptions. He then proceeds to examine the suitability of sections of the plan where he has special knowledge. He throws doubts on the ability of the farmers to reach certain targets of production set by the planners, and bases his scepticism on the resistance of small-scale producers of export crops to the adoption of improved techniques, and the character of the Ceylonese peasant which will make it difficult to raise rice yields. Farmer warns that if the plan fails to reach its objectives the cause may be found in non-economic factors, which include political instability.

REFERENCES

ALEXANDER, J. W., 1963, *Economic geography*, New Jersey.

ALEXANDER, L. M., 1957, *World political patterns*, Chicago.

ANDERSON, C. W., 1967, *Politics and economic change in Latin America*, New York.

BAER, W., 1964, 'Regional inequality and economic growth in Brazil', *Economic Development and Cultural Change*, **12**, pp. 268–85.

BETTELHEIM, C., 1963, 'India's third five year plan', *Pacific Viewpoint*, **4**, pp. 139–55.

BLACK, L. D., 1963, 'U.S. economic aid to Africa' (Abstract), *Annals*, Association of American Geographers, **53**, pp. 579–80.

BLIJ, H. J. de, 1967, *Systematic political geography*, New York.

BLOUET, B. W., 1965, 'Malta's first five year plan', *Geography*, **50**, pp. 73–5.

BOWDEN, P. J., 1965, 'Regional problems and policies in the Northeast of England', *J. of Industrial Economics* (Supplement), pp. 20–39.

BUCKHOLTS, P., 1966, *Political geography*, New York.

CAMM, J. C. R., 1967, 'The Queensland Agricultural Land Purchase Act 1894 and rural settlement', *Australian Geographer*, **10**, pp. 263–74.

CLARK, R. G., 1967, *U.S. non-military aid to Africa*, unpublished paper, University of Melbourne.

COMMONWEALTH BUREAU OF CENSUS AND STATISTICS, 1966, *Year Book, Australia* 1966, **52**, Canberra.

COPPOCK, J. T., *1966*, 'Agricultural developments in Nigeria', *J. of Tropical Geography*, **23**, pp. 1–18.

COURTENAY, P. P., 1961, 'International tin restriction and its effects on the Malayan tin mining industry', *Geography*, **46**, pp. 223–31.

CULBERTSON, W. S., 1924, 'Raw materials and foodstuffs in the commercial policies of nations', *Annals*, American Academy of Political and Social

Science, **112**, pp. 1–133.

DANIEL, J. B. McI., 1966, 'Some government measures to improve African agriculture in Swaziland', *Geogr. J.*, **132**, pp. 506–14.

DAVEAU, S., 1959, *Les régions frontalières de la montagne Jurassienne*, Paris.

EDWARDS, K. C., 1964, 'The new towns of Britain', *Geography*, **49**, pp. 279–85.

FARMER, B. H., 1961, 'The Ceylon ten year plan 1959–68', *Pacific Viewpoint*, **2**, pp. 123–36.

FERRER, A., 1967, *The Argentine economy*, Los Angeles.

FINANCIAL MAIL, 14 July 1967, *1961–6: the fabulous years*, Johannesburg.

FISHER, Sir G., 1967, 'Overseas investment in the Australian mining industry', Presidential address, The Australasian Institute of Mining and Metallurgy, *Proceedings*, 223, Melbourne.

FITZPATRICK, B. and Wheelwright, E. L., 1965, *The highest bidder*, Melbourne.

FREEBERNE, M., 1962, 'Natural calamities in China 1949–61', *Pacific Viewpoint*, **3**, pp. 33–72.

FREEMAN, T. W., 1967, *Geography and planning*, London.

GENERAL AGREEMENT ON TARIFFS AND TRADE, 1958, *Commercial policy 1957*, Geneva.

GERASIMOV, I. P., Armand, D. L., and Preobrazhenskiy, V. S., 1964, 'Natural resources of the Soviet Union, their study and utilisation', *Soviet Geography*, **5**, **8**, pp. 3–14.

GOODWIN, W., 1962, 'Relocation of the British motor industry', *Professional Geographer*, **14**, pp. 4–8.

GROTEWOLD, A., and Sublett, M. D., 1967, 'The effect of import restrictions on landuse: the United Kingdom compared with West Germany', *Economic Geography*, **43**, pp. 64–70.

HAMILTON, F. E. I., 1964, 'Location factors in the Yugoslav iron and steel industry', *Economic Geography*, **40**, pp. 46–64.

HIGHSMITH, R. M., and Jensen, J. G., 1963, *Geography of commodity production*, New York.

HUDSON, S. C., 1961, 'Role of commodity agreements in international trade', *J. of Agricultural Economics*, **14**, 507–30.

HUMPHRYS, G., 1965, 'Economic change and regional problems in industrial South Wales', *J. of Industrial Economics* (Supplement), pp. 40–53.

JACKSON, J. N., 1966, *Surveys for town and country planning*, London.

LARGE, D. C., 1957, 'Cotton in the San Joaquin Valley: a study of government in agriculture', *Geogr. Rev.*, **47**, pp. 365–80.

LEE, T. R., 1966, *The distribution and mobility of Italians in Melbourne*. Unpublished thesis, University of Melbourne.

LEFTWICH, R. H., 1966, 'Exchange rate policies, balance of payments, and trade restrictions in Chile', *Economic Development and Cultural Change*, **14**, pp. 400–13.

LEWIS, J. N., 1967, 'Agricultural price policies', in *Agriculture in the Australian economy*, Ed. Williams, D. B., Sydney.

LINGE, G. J. R., 1967, 'Governments and the location of secondary industry in Australia', *Economic Geography*, **43**, pp. 43–63.

MCCRONE, G., 1965, 'Next steps in regional planning', *J. of Industrial Economics* (Supplement), pp. 115–30.

MCKNIGHT, T., 1967, 'Industrial location in South Australia', *Australian Geographical Studies*, **5**, pp. 50–72.

MACLENNAN, M. C., 1965, 'Regional planning in France', *J. of Industrial Economics* (Supplement), pp. 62–75.

MALEK, H., 1966, 'Après la réforme agraire Iranienne', *Annales de Géographie*, **409**, 268–85.

MINSHULL, R., 1967, *Regional geography*, London.

MOES, J. E., 1966, 'Foreign exchange policy and economic union in Central Africa', *Economic Development and Cultural Change*, **14**, pp. 471–83.

NAYLON, J., 1967, 'Irrigation and internal colonisation in Spain', *Geogr. J.*, **133**, pp. 178–91.

OSALA, E. M., 1967, 'Some current aspects of international commodity policy', *J. of Agric. Economics*, **18**, pp. 27-46.

PARKS, A. C., 1965, 'The Atlantic Provinces of Canada', *J. of Industrial Economics* (Supplement), pp. 76–87.

PINCUS, J., 1967, *Trade, aid and development*, New York.

POUNDS, N. J. G., 1963, *Political geography*, New York.

PRESCOTT, J. R. V., 1959, 'Migrant labour in the Central African Federation', *Geogr. Rev.*, **49**, pp. 424–7.

PRESCOTT, J. R. V., 1961, 'Overpopulation and overstocking in the Native areas of Matabeleland', *Geogr. J.*, **127**, pp. 212–24.

PRESCOTT, J. R. V., 1966, 'Resources, policy and development in West and North Central Africa', *Northern Geographical Essays*, Ed. House, J. W., pp. 292–309, Newcastle upon Tyne.

PRESCOTT, J. R. V., 1967, *The geography of frontiers and boundaries*, London.

PRESTON, Y., 1967, 'Economic policy: the art of imperfect compromises', *Financial Review*, 8 December 1967, p. 3, Sydney.

REEVE, W. H., 1960, 'Progress and geographical significance of the Kariba Dam', *Geogr. J.*, **126**, pp. 140–6.

RICHARDS, A. I., 1958, 'A changing pattern of agriculture in East Africa: the Bemba of Northern Rhodesia', *Geogr. J.*, **124**, pp. 302–14.

RUTHERFORD, J., 1963, 'Group irrigation schemes and integrated uses of land in the southern Murray-Darling Basin', *Australian Geographical Studies*, **1**, pp. 65–83.

SAUSHKIN, Y. S., Grishin, G. T., Stepanov, M. N., Ivanov, S. I., and Novikov, V. P., 1964, 'An approach to the economic-geographic modeling of regional territorial-production complexes', *Soviet Geography*, **5**, **10**, pp. 19–32.

SCHEEL, W., 1963, 'Die Politische und Wirtshaftliche Bedeutung der Entwicklungshiffe für die Lander Afrikas', *Die Erde*, **94**, pp. 182–90.

SCHICKELE, R., 1954, *Agricultural policy: farm programs and national welfare*, Lincoln.

SCOTT, P., 1954, 'Migrant labour in Southern Rhodesia' and 'The role of Northern Rhodesia in African labour migrations', *Geogr. Rev.*, **44**, pp. 29–48 and 432–4.

SEAWALL, F., 1961, 'Recent developments in Mexican sulphur production', *J. of Tropical Geography*, **15**, pp. 39–45.

SMITH, R. H. T., 1964, 'The development and function of transport routes in southern New South Wales, 1860–1930', *Australian Geographical Studies*, **2**, pp. 47–65.

SYMONS, L., 1967, 'Russia and the Third World', paper presented to *Australian and New Zealand Association for the Advancement of Science*, Section P., 39th Congress, Melbourne.

THOMAN, R. S. and Conkling, E. C., 1967, *Geography of international trade*, New Jersey.

THOMAS, F. H., 1962, 'Economically distressed areas and the role of the academic geographer', *Professional Geographer*, **14**, pp. 12–15.

TOKAREV, S., and Alampiyev, P., 1962, 'Problems of improving the territorial organisation of the national economy and economic regionalisation', *Soviet Geography*, **3**, **1**, pp. 39–48.

TOWLE, L. W., 1956, *International trade and commercial policy*, New York.

UNITED NATIONS, 1949, *National and international measures for full employment*, New York.

UNITED NATIONS, 1949a, *Maintenance of full employment*, New York.

UNITED NATIONS, 1963, *Permanent sovereignty over natural wealth and resources*, New York.

UNITED STATES DEPARTMENT OF AGRICULTURE, 1964, 'Agricultural policies of foreign governments including trade policies affecting agriculture', *Agricultural Handbook*, **132**, Washington.

WEIGERT, H. W., 1957, *Principles of political geography*, New York.

WILLIAMSON, J. G., 1965, 'Regional inequality and the process of national development', *Economic Development and Cultural Change*, **13**, part 2.

WILSON, T., 1965, 'Papers on regional development', *J. of Industrial Economics* (Supplement), pp. vii–xiv.

WOYTINSKY, W. S., and Woytinsky, E. S., 1955, *World commerce and governments*, New York.

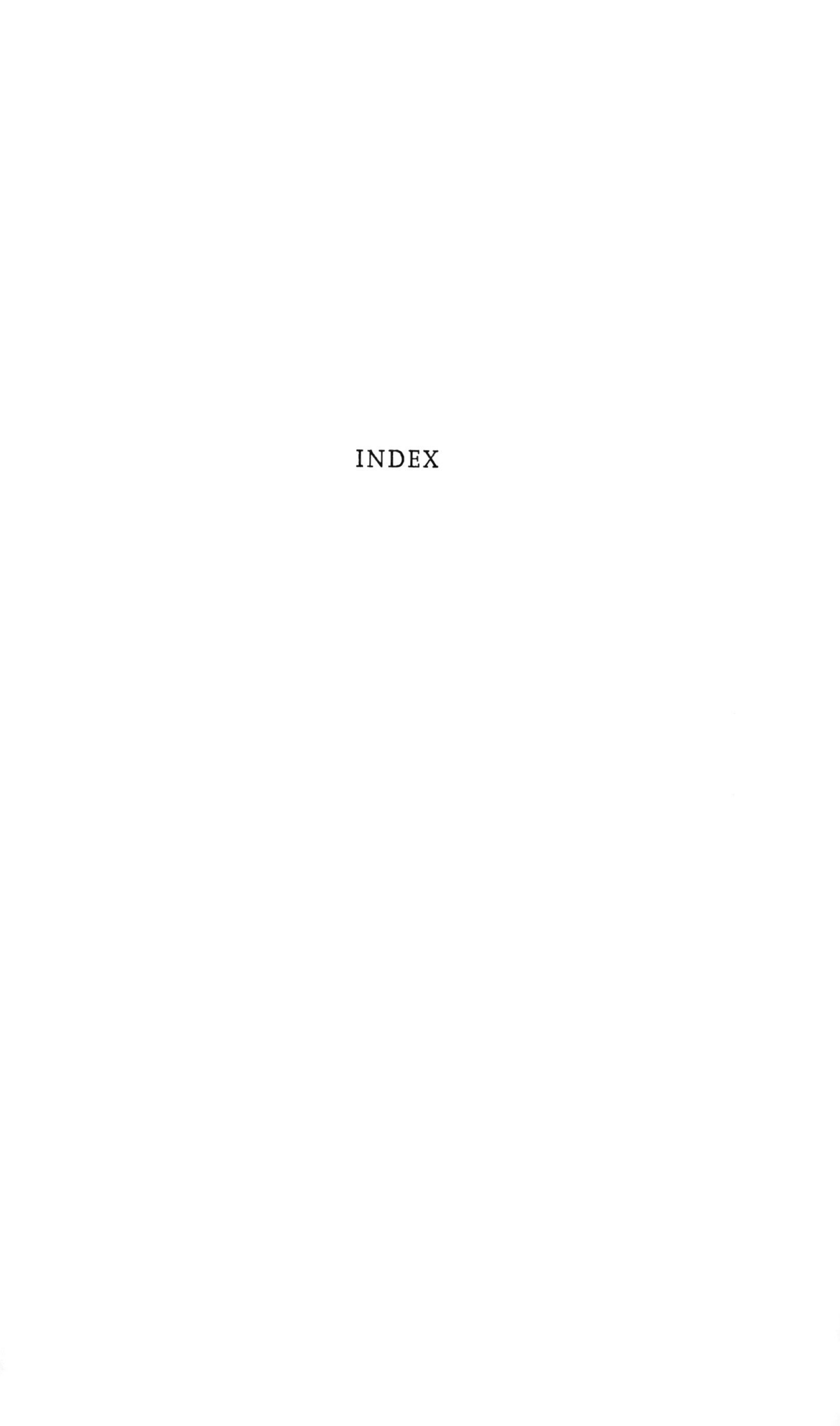

INDEX

Printed and bound by CPI Group (UK) Ltd, Croydon, CR0 4YY
22/10/2024
01777620-0017